Springer Theses

Recognizing Outstanding Ph.D. Research

Aims and Scope

The series "Springer Theses" brings together a selection of the very best Ph.D. theses from around the world and across the physical sciences. Nominated and endorsed by two recognized specialists, each published volume has been selected for its scientific excellence and the high impact of its contents for the pertinent field of research. For greater accessibility to non-specialists, the published versions include an extended introduction, as well as a foreword by the student's supervisor explaining the special relevance of the work for the field. As a whole, the series will provide a valuable resource both for newcomers to the research fields described, and for other scientists seeking detailed background information on special questions. Finally, it provides an accredited documentation of the valuable contributions made by today's younger generation of scientists.

Theses are accepted into the series by invited nomination only and must fulfill all of the following criteria

- They must be written in good English.
- The topic should fall within the confines of Chemistry, Physics, Earth Sciences, Engineering and related interdisciplinary fields such as Materials, Nanoscience, Chemical Engineering, Complex Systems and Biophysics.
- The work reported in the thesis must represent a significant scientific advance.
- If the thesis includes previously published material, permission to reproduce this must be gained from the respective copyright holder.
- They must have been examined and passed during the 12 months prior to nomination.
- Each thesis should include a foreword by the supervisor outlining the significance of its content.
- The theses should have a clearly defined structure including an introduction accessible to scientists not expert in that particular field.

More information about this series at http://www.springer.com/series/8790

Fumika Nagasawa

Studies on the Plasmon-Induced Photoexcitation Processes of Molecules on Metal Surfaces

Doctoral Thesis accepted by
Hokkaido University, Sapporo, Japan

Springer

Author
Dr. Fumika Nagasawa
Graduate School of Chemical Sciences and Engineering
Hokkaido University
Sapporo
Japan

Supervisor
Prof. Kei Murakoshi
Graduate School of Chemical Sciences and Engineering
Hokkaido University
Sapporo
Japan

ISSN 2190-5053 ISSN 2190-5061 (electronic)
Springer Theses
ISBN 978-4-431-56577-2 ISBN 978-4-431-56579-6 (eBook)
DOI 10.1007/978-4-431-56579-6

Library of Congress Control Number: 2017946643

Printed on acid-free paper

This Springer imprint is published by Springer Nature
The registered company is Springer Japan KK
The registered company address is: Chiyoda First Bldg. East, 3-8-1 Nishi-Kanda, Chiyoda-ku, Tokyo 101-0065, Japan

Supervisor's Foreword

The photoexcitation process is the basis for photochemical/physical technologies. If one can achieve desirable control of photoexcitation processes, technologies for energy, information, and sensing can be improved drastically. To control the process, we must change or modify the light–matter interaction. Although we know the limitation of the interaction, which is determined by the selection rules of electronic excitation under "normal light" illumination. Recently, "confined light" in metal nanostructures known as localized surface plasmon has attracted much attention. Such small light can be utilized as being beyond the limit of interaction with materials. Strong interaction between a confined light field and materials results in the formation of a novel absorption band. Despite expectations of using systems showing a strong interaction, the methodology to create those systems, as well as highly sensitive observation of the exotic behaviour of nanosystems, has not yet been established.

In this thesis, Dr. Fumika Nagasawa reveals the characteristic properties of the photoexcitation process by excitation of the molecules on or near metal nanostructures. First, she fabricated a highly ordered nanostructure and evaluated the polarization of Raman scattering photons from the structure (Chap. 2). Simultaneous measurement on conductance and Raman scattering at a single molecule junction was carried out to determine the characteristics of the electronic excitation at the interface between metal and molecule (Chap. 3). Photoexcitation processes were evaluated by using a single-walled carbon nanotube in the metal nanogap in the electrochemical environment (Chap. 4). For more active modulation of the photoexcitation process, a hybridized system of molecule and metal was prepared in order to achieve a strong coupling regime between molecular excitons and plasmons in metal nanostructures (Chap. 5). Finally, strong coupling states were controlled by adopting electrochemical potential tuning (Chap. 6).

The findings presented in this thesis will offer a novel route for molecule photoexcitation. The methodology and principles for creating a system showing exotic light–matter interaction were established, and desirable control of the system was also achieved. Novel insight into the photoexcitation process within a hybridized system may contribute to drastic changes in photochemical/physical technologies in the future.

Sapporo, Japan

Prof. Kei Murakoshi

Parts of this thesis have been published in the following journal articles and book:

1. Hiro Minamimoto, Fumiya Kato, Fumika Nagasawa, Mai, Takase, and Kei Murakoshi, "Electrochemical Control of Strong Coupling States between Localized Surface Plasmons and Molecule Excitons for Raman Enhancement", *Faraday Discussion*, DOI: 10.1039/C7FD00126F (2017).
2. M. Takase, F. Nagasawa, H. Nabika, and K. Murakoshi, "Depolarization of Surface-Enhanced Raman Scattering Photons from a Small Number of Molecules on Metal Surfaces", in *Frontiers of Surface-Enhanced Raman Scattering: Single-Nanoparticles and Single Cells*, edited by Y. Ozaki, K. Kneipp and R. R. Aroca (John Wiley & Sons, New York, 2014), p. 89–106.
3. F. Nagasawa, M. Takase, and K. Murakoshi, "Raman Enhancement via Polariton States Produced by Strong Coupling between a Localized Surface Plasmon and Dye Excitons at Metal Nanogaps", *J. Phys. Chem. Lett.*, **5**, 14–19 (2014).
4. F. Nagasawa, M. Takase, H. Nabika, and K. Murakoshi, "Single Molecule Surface-Enhanced Raman Scattering as a Probe for Adsorption Dynamics on Metal Surfaces", in *Vibrational Spectroscopy at Electrified Interfaces*, edited by C. Korzeniewski, B. Braunschweig and A. Wieckowski (John Wiley & Sons, New York, 2013), p. 220–237.
5. T. Konishi, M. Kiguchi, M. Takase, F. Nagasawa, H. Nabika, K. Ikeda, K. Uosaki, K. Ueno, H. Misawa, and K. Murakoshi, "Single Molecule Dynamics at a Mechanically Controllable Break Junction in Solution at Room Temperature", *J. Am. Chem. Soc.*, **135**, 1009–1014 (2013).
6. F. Nagasawa, M. Takase, H. Nabika, and K. Murakoshi, "Polarization Characteristics of Surface-Enhanced Raman Scattering from a Small Number of Molecules at the Gap of a Metal Nano-Dimer", *Chem. Commun.*, **47**, 4514–4516 (2011).
7. F. Nagasawa, M. Takase, H. Nabika, and K. Murakoshi, "Characteristic Surface-Enhanced Raman Scattering from a Small Number of Molecules in an Anisotropic Electromagnetic Field at Metal Nano-Gap", *Trans. Mater. Res. Soc. Jpn.*, **35**, 279–282 (2010).

Acknowledgements

I would like to express my sincere gratitude to Prof. Kei Murakoshi for his discussions and continuous encouragement throughout this study.

I also would like to express deep appreciation to Prof. Koichiro Ishimori, Prof. Noboru Kitamura, Prof. Tamotsu Inabe, Prof. Jyunji Nishii, and Prof. Toshihiro Shimada for their important discussions and very useful suggestions.

I am grateful to Dr. Satoshi Yasuda, Prof. Manabu Kiguchi (Tokyo Institute of Technology) and Dr. Hideki Nabika (Yamagata University), Dr. Katsuyoshi Ikeda for their valuable discussions and helpful advice. Moreover, I am grateful to Prof. Hiroaki Misawa, Dr. Ueno Kosei, Prof. Yasuyuki Tsuboi, and Dr. Tamitake Ito for their valuable comments and discussions about the experiments. I wish to express my thanks to Dr. Tatsuya Konishi, Dr. Baku Takimoto and Dr. Toshinori Motegi for their helpful suggestions on my studies. I am sincerely grateful to Dr. Mai Takase for teaching me consistent appreciation of myself and those around me in my perpetual journey for self-improvement. I am grateful to all of the colleagues at the Laboratory of Material Chemistry and the Laboratory of Physical Chemistry, Department of Chemistry, Faculty of Science, Hokkaido University, for their indispensable comments and for creating a cheerful, productive environment in the laboratory

Finally, I would like to express my greatest gratitude to my parents and my sister for their hearty support and encouragement throughout these years, without which I would not have been able to achieve this work.

Fumika Nagasawa

Contents

Chapter 1
General Introduction

The photoexcitation process is the basis for photochemical/physical technologies, such as photoenergy conversion, artificial photosynthesis and light-emitting diodes. If we are able to control the photoexcitation processes, these technologies could be drastically improved. The interaction between light and molecules is extremely restricted due to the intrinsic electronic structure of material. Moreover, the limitation of the maximum strength of interaction in the optimal electronic structure is well known. To address this limitation, I applied surface plasmons (SPs) for the photoexcitation of materials. SPs represent the collective oscillation of free electrons at the surface of metal. As a result, a highly confined, anisotropic electromagnetic (EM) field localised near the metal surface can be induced by light illumination. A metal's EM field can be affected by materials located on its surface. The specified direction of the EM field can be used to evaluate the transition of the electronic state of the target material. Moreover, a specified energy in an SP state may satisfy the resonant condition with the molecular excitonic energy. It is also expected to generate and control the coupled state between the molecular excitons and SP energy.

1.1 Surface Plasmon Resonance for Control of the Photon Field

It is suitable to use SPs for confinement of the photon field near the metal surface. The highly localised EM field near the metal surface induces enhancement phenomena, such as surface-enhanced fluorescence (SEF) and surface-enhanced Raman scattering (SERS). To control the energy, wavenumber and anisotropy of the EM field, systems combining metals and dielectric materials for SP resonance (SPR) have been developed.

F. Nagasawa, *Studies on the Plasmon-Induced Photoexcitation Processes of Molecules on Metal Surfaces*, Springer Theses,
DOI 10.1007/978-4-431-56579-6_1

In the state of the free movement of ions or electrons (plasma), the term *plasmon* describes the collective oscillation of free ions or electrons. In the metal, rich free electrons oscillate collectively (plasmon). The collective oscillation propagating in the bulk metal is a longitudinal wave that forbids interaction with the transverse wave of the EM field. At the surface of the metal, SP propagating at the metal/dielectric interface leaks into the dielectric component as an evanescent field. The EM field of the evanescent field has the component of a transverse wave. If the surface becomes extremely confined to an area of around a few hundred nm, such as in metal nanoparticles, the EM field is quantised.

For the excitation of the SPR, the Otto configuration [1] and Kretschmann configuration [2, 3] were reported in 1968. The Otto configuration was experimentally realised by using a prism on top of a metal surface with a small air gap, as depicted schematically in Fig. 1.1a. The prism was made from a dielectric medium with a high refractive index for increments of the wave vector. The evanescent field caused by the total internal reflection on the prism surface coupled to the evanescent field of the SP on the metal surface. The appropriate air gap distance caused excitation of SP by light illumination. In contrast, the Kretschmann configuration is depicted schematically in Fig. 1.1b, representing the interchanged configuration of metal and air. In the Kretschmann configuration, energy is exchanged between the evanescent wave and SP on the air side. The Otto and Kretschmann configurations are referred to collectively as attenuated total reflection (ATR).

To achieve more effective excitation for propagating SP without prisms, diffraction grating [4, 5], subwavelength hole arrays [6, 7] and the bullseye structure [8] have been adopted. Here, the incident light wave from the diffraction can couple to the SP. As a group, these structures are called plasmonic crystals, and they are similar to photonic crystals. The SP of the thin film in diffraction grating couples to the reverse side of metal film, providing the ability to propagate the coupled plasmon far away [9]. In this mode, the propagation length is long and the loss is small (resulting in the term *long-range plasmon* [10]). This mode was first reported by Fukui et al. on a thin flat film layer [11].

In a similar fashion, nanosized metal particles can interact with light; this phenomenon is called localised SPR (LSPR). LSPR is induced in many structures, such

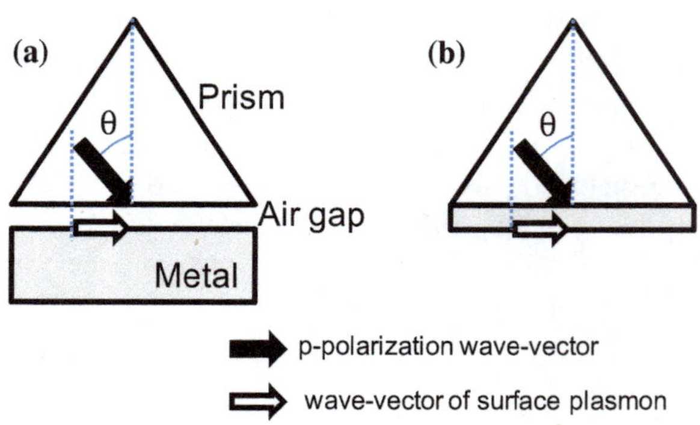

Fig. 1.1 **a** Otto configuration, **b** Kretschmann configuration

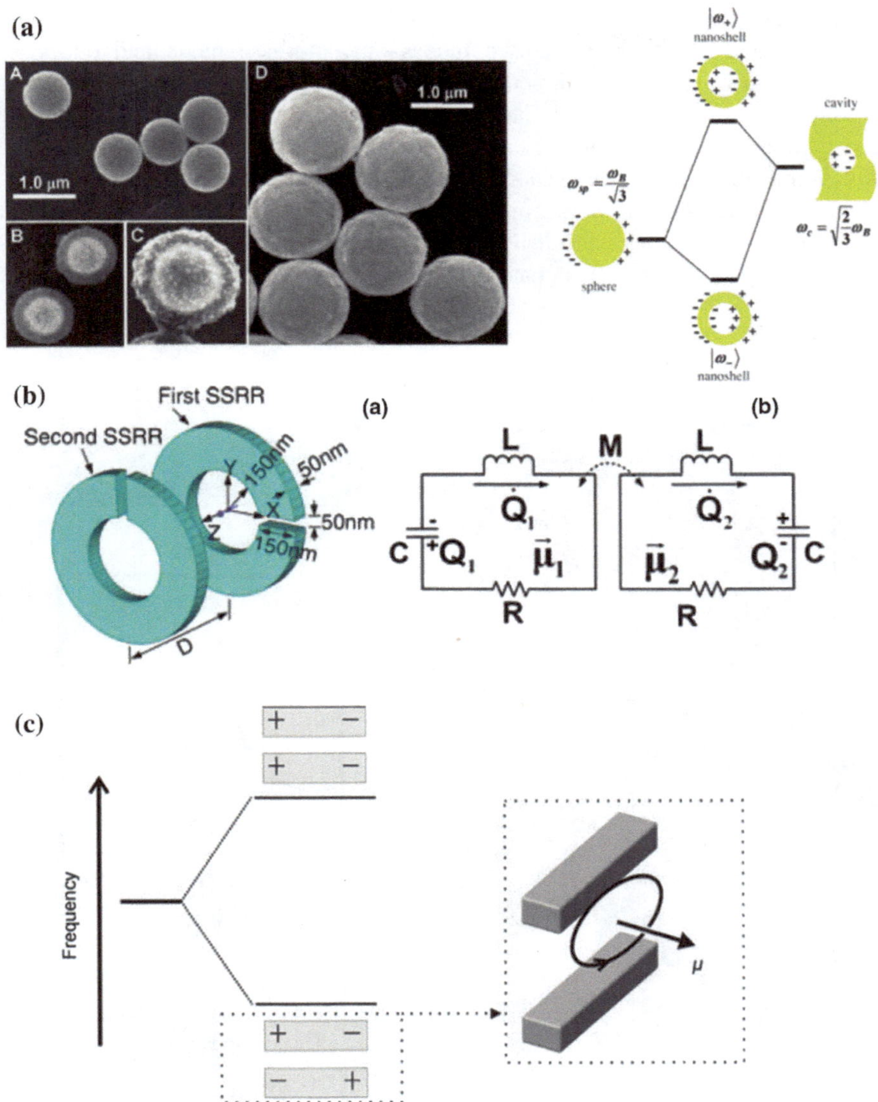

Fig. 1.2 Plasmonic coupling: **a** nanoshell [20], **b** metamaterials [18], **c** magnetic atoms [19]

as nanospheres, nanodimers, nanorods and nanoshells [12]. The EM field of LSPR is induced in the proximity of the metal surface; in other words, the incident EM field is confined and enhanced near this surface. The polarisation of plasmon resonance, which oscillates in the same frequency as the incident light, can radiate the propagating wave to infinity. The appropriate proximity results in plasmonic coupling between LSPR and LSPR, SPR and SPR and SPR and LSPR. In the condition of the same resonant frequency and strong coupling between two plasmonic modes, these mode cause hybridisation, a function that is called *normal mode splitting*. Here, the interference between two modes—the in-phase mode and out-phase mode—is measured. A nanodimer structure organised from a nanoparticle pair is the simplest case. The in-phase plasmon mode is easily excited by the light and the EM field is strongly confined in the gap. The enhanced EM field shows the same direction as the excited polarisation, which is the parallel to the dimer's long axis. Figure 1.2 shows the plasmon coupling materials. If we change one nanoparticle of the dimer to a cavity, a nanoshell structure [13–17] is fabricated. A more asymmetric structure provides metamaterial [18] (Fig. 1.2b) and magnetic atoms arise

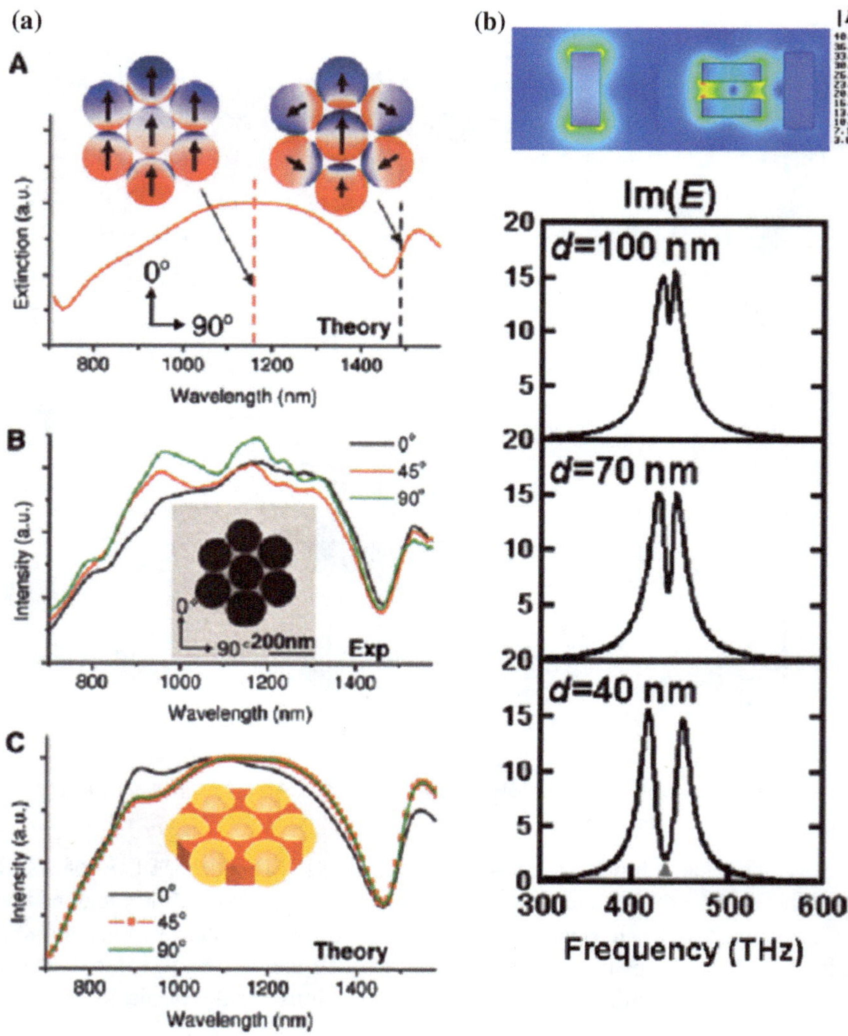

Fig. 1.3 Fano resonance [22] and plasmon-induced transparency in metamaterial [23]

[19] (Fig. 1.2c). The EM-induced transparency and Fano resonance occur when one plasmonic mode has high loss and another has low loss.

The first observation of Fano resonance was Wood's anomaly [21]. This structure showed the interference between incident light and reradiation from the SP. The coupling between plasmons was measured in the heptamer of Au nanoparticles [22]. The EM-induced transparency was also caused by plasmonic mode coupling [23]. Figure 1.3b shows the EM-induced transparency at the resonant frequency. Here, energy transfer between plasmons is effectively induced. The fabrication technique opens the wide research avenue to control and modulate the range of photonic and optical properties.

1.2 Interaction Between Plasmons and Molecules: Active Plasmonics

The interaction between metal and molecule needs to be controlled to enhance the signal modulation. There is a perturbation of localised SPs (LSPs) at the gap structure as a plasmonic source. Below, the fascinating characteristics of these structures are discussed relating to an intrinsic feature of interaction, termed *active plasmonics*.

1.2.1 Use of Plasmons as a Confined Photon Field (Electromagnetic Enhancement)

We can use plasmonic structures as a confined photon source, and this enhances the EM field. The relevant features are known as the EM effect, Purcell effect [24], plasmon resonance energy transfer (PRET) [25] and plasmophore [26, 27]. These basic mechanisms are used for applications like plasmon-enhanced solar cells (dye sensitised solar cells [28, 29], silicon light-harvesting devices [30, 31]), organic light-emitting devices [32]. Plasmon resonance is concerned with the excitation and radiation processes. In the excitation process, the plasmonic resonance induces the EM effect of incident light. This effect is considered as the same mechanism by which strong light irradiates to molecules when weak light irradiates to plasmonic structures. Another process shows the enhancement of radiation efficiency. As is the case of Fermi's golden rules, each molecule has an intrinsic emission rate in nature. However, if the surrounding environment is modulated, the emission rate is affected. When molecules are situated in the plasmonic cavity, changes occur in the radiative emission rate, nonradiative emission and hole–electron coupling. The radiative emission of photons occurs in the appropriate proximity. The merits of using plasmonic structures as a photon source are the enhanced excitation efficiency and quantum efficiency for the suppression of nonradiative processes.

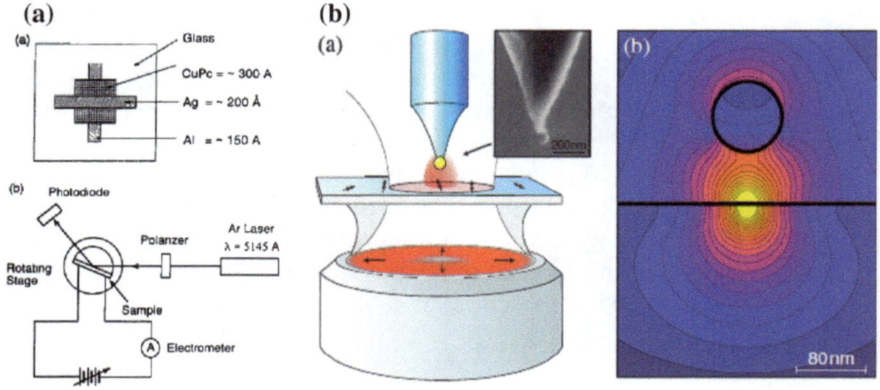

Fig. 1.4 **a** Plasmonic enhanced organic solar cell [28] and **b** plasmonic enhanced fluorescence [33]

The pioneering observation of the plasmonic solar cell is shown in Fig. 1.4a [28]. This organic solar cell provided an enhanced photocurrent via propagating SPR on a flat metal surface in the Kretschmann configuration. The observed enhancement is due to the increase in photocarrier generation, which results from the enhanced light absorption under SPP excitation. A similar enhancement mechanism is measured in the plasmon-enhanced fluorescence shown in Fig. 1.4b [33]. The gap distance between the tip-embedded metal nanoparticles and fluorescence molecules affects the fluorescence intensity. However, quenching occurs in the nearest proximity, and the maximum fluorescence intensity is measured at the appropriate gap distance. Other measurements and applications, such as plasmon-enhanced Si solar cells [30, 31, 34–36], plasmonic light-emitting diodes [37–39] and plasmonic lasers [40–43], are developed using the EM enhancement mechanism.

1.2.2 Use of Plasmons for Generating Electron–hole Pairs via Plasmon Decay

In contrast to the EM enhancement in Sect. 1.2.1, another important property of optical antennas is their tendency to generate energetic or ‘hot’ electron–hole pairs by plasmon decay [44–49]. Light is absorbed to form electron–hole pairs. This process of hot-electron generation has been shown to participate in photochemical reactions on the surface of noble metal nanoparticles [50–54]. The use of hot electrons is applicable for photoenergy conversion as shown in Fig. 1.5. Active plasmon nanoparticles perform the photocurrent generation as sensitisers (Fig. 1.6) [55–58]; this photoenergy conversion technique is expected to improve conversion efficiency and extend to the absorption wavelength region in longer wavelengths.

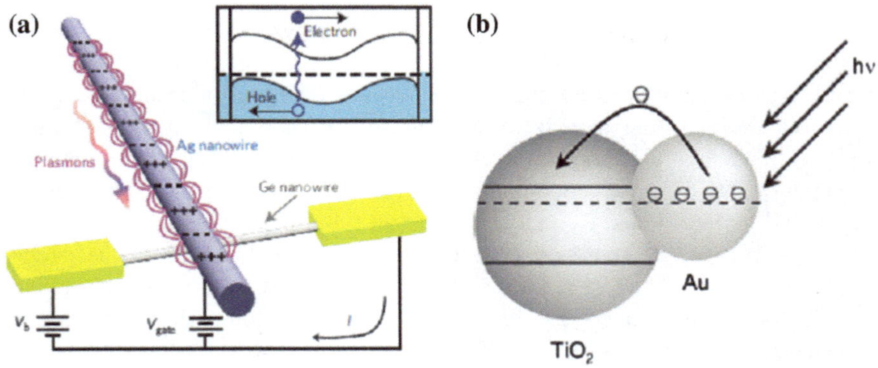

Fig. 1.5 **a** Propagating plasmon-induced photocurrent measurement [77], **b** photocurrent measurement using plasmonic Au nanoparticles [55]

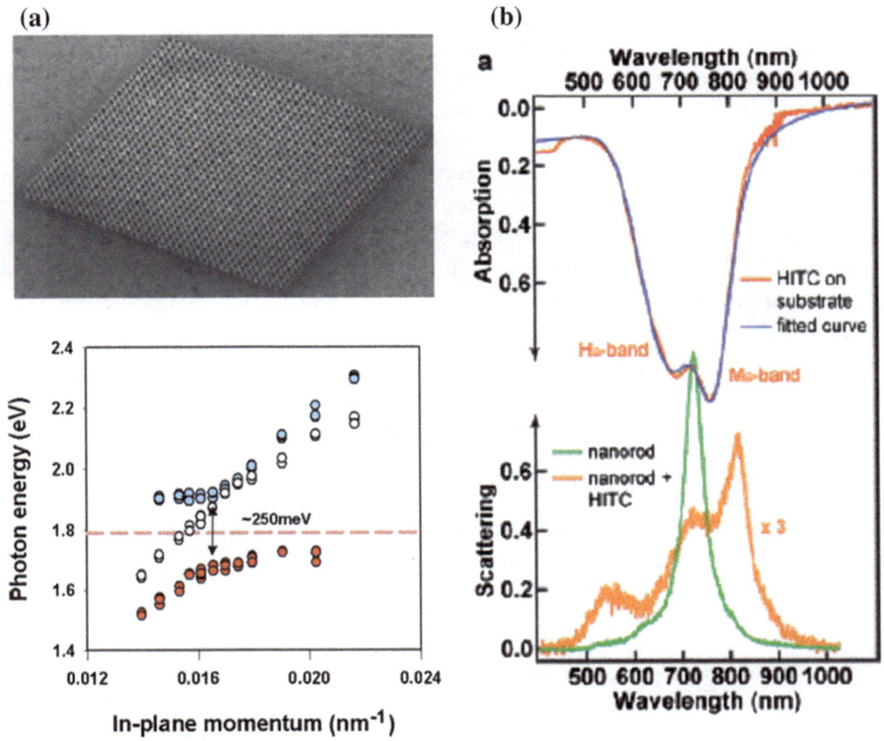

Fig. 1.6 Strong coupling between organic dye molecule and **a** propagating nanohole surface plasmon [63] and **b** localised surface plasmon [71]

1.2.3 Formation of the Hybridised State

Intensive investigations of the interactions between the electronic states in organic molecules and plasmons have been undertaken. We know that organic dye molecules can have high oscillator strengths that form strong-coupling regimes at room temperature [59–67]. The coupling strength can be controlled via the plasmon energy and its linewidths. Recently, it has been found that hybrid states modulate optical absorption and emission properties [59–61], the lifetimes of the excited states [62] and the chemical reactivity [66] of the systems. Although many reports have observed strong coupling in the field of SPs, application of the LSPR remains a challenging subject [68–72].

1.3 Surface-enhanced Raman Scattering

Plasmon resonance is an extremely useful technique for photoexcitation and photoinduction processes. In this resonant condition, it is important to analyse the molecules situated on the metal surface. The response of molecules under the vibrational spectroscopy technique is enhanced using SPR via surface-enhanced phenomena. Raman scattering is a useful technique for accurately identifying molecules; moreover, it can be performed in situ and is nondestructive. It is a good design for the in situ Raman measurement of the interface between metal and molecules for use in the plasmonic enhancement field. Raman scattering is a two-photon process that includes information about excitation and scattering. The benefit of this double resonance condition is that it provides a strongly enhanced photon field at the edge of the metal nanostructure. This shows promise as a technique for extracting information about the electronic states of the target molecules on the metal surface. Furthermore, it is known that the intensity of Raman scattering is also enhanced by the resonance of the electronic excitation in the electronic states at the interface between molecules and metals. Enhanced Raman signals via various mechanisms can provide information on chemisorbed and physisorbed molecules on metal surfaces.

1.3.1 Normal Raman Scattering

The polarisability of the metal nanodimer array shows intense measurement. Our signal enhancement illustrates the minimum survey at the intrinsic enhancement.

The Raman polarisability of molecules is given by the following:

$$\alpha = \alpha_0 + \left(\frac{\partial \alpha}{\partial Q}\right)_0 Q, \tag{1.1}$$

$$Q = Q_0 \cos 2\pi \nu t, \tag{1.2}$$

$$E = E_0 \cos 2\pi \nu t. \tag{1.3}$$

The induced dipole moment is

$$\mu = \alpha E, \tag{1.4}$$

$$\mu = \alpha_0 E_0 \cos 2\pi \nu_i t + \frac{1}{2}\left(\frac{\partial \alpha}{\partial Q}\right)_0 Q E_0 [\cos 2\pi(\nu_i + \nu)t + \cos 2\pi(\nu_i - \nu)t]. \tag{1.5}$$

The first term indicates the Rayleigh scattering, while the second term corresponds to Raman scattering. We can obtain the Raman-scattered light if the derived polarisability becomes nonzero as

$$\left(\frac{\partial \alpha}{\partial Q}\right)_0 \neq 0. \tag{1.6}$$

This correlation is satisfied when the polarisability changes with the vibration. Polarisability is derived by the polarisation tensor in the space coordination as follows:

$$\alpha = \begin{pmatrix} \alpha_{xx} & \alpha_{xy} & \alpha_{xz} \\ \alpha_{yx} & \alpha_{yy} & \alpha_{yz} \\ \alpha_{zx} & \alpha_{zy} & \alpha_{zz} \end{pmatrix}. \tag{1.7}$$

The totally symmetric mode has a diagonal element, while the nontotally symmetric mode has a nondiagonal element. The EM unit vector is represented by the tensor, which can be used to denote the polarisation configuration as follows:

$$e_i = \begin{pmatrix} x \\ y \\ z \end{pmatrix}, \quad e_s = \begin{pmatrix} x \\ y \\ z \end{pmatrix}. \tag{1.8}$$

We can deduce the Raman intensity ($I_s R^2$) as

$$I_s R^2 = \frac{\omega_s^4}{c^4} (\mathbf{e}_s \alpha \mathbf{e}_i)^2 I_i. \tag{1.9}$$

I_i is the intensity of the incident planar EM wave, and ω_s is the angular frequency of scattering photons. Both the incident polarisation and scattering polarisation are related to the intensity of the Raman scattering.

If polarised Raman measurement is performed, the intensity ratio of the Raman spectrum changes in each orientation to the intrinsic polarisation configuration. For example, in the backscattering configuration that has 0° between the incident and scattering propagating directions, when we use the parallel incident and scattering polarisation to the x-direction, the intensity ratio at each polarisation in each polarisation angle, the tensor element α_{xx} engages the Raman scattering intensity. The system of the single crystal of molecules fixed at the specific orientation in the space enables us to measure the tensor elements. However, the liquid or the solution in which molecules are oriented randomly shows random tensor elements. If we consider the depolarisation ratio, we can analyse whether the totally symmetric mode or nontotally symmetric mode is present. The depolarisation ratio ρ is defined by the ratio of the Raman intensity of perpendicular polarisation to parallel polarisation. The totally symmetric mode shows $0 \leq \rho < 0.75$ and the nontotally symmetric mode shows $\rho = 0.75$.

1.3.2 The Electromagnetic Effect of Surface-enhanced Raman Scattering

If we use metal nanoparticles, the Raman signal is strongly enhanced by the induced LSPR. Surface-enhanced Raman scattering intensity is determined as follows:

$$I_s R^2 = \frac{8\pi(\omega \pm \omega_{IF})^4}{9c^4} I_L \sum_{\sigma,\rho} |\alpha_{\sigma\rho}|^2. \tag{1.10}$$

I_L is the intensity of the incident planar EM wave, ω_{IF} is the angular frequency of scattering photons and **e** is the electronic vector. The intensity of the Raman scattering is determined by the both the incident and scattering EM field, like the normal Raman process. Intense Raman enhancement originates from the twofold EM enhancement mechanism that arises from the coupling of plasma resonance with both incident and Raman-scattered light [73–76]. First, the light shines onto the metal nanoparticles, and LSPR absorption will occur. Then, LSPR is excited, and Raman scattering radiation will occur in the region of the intense Raman enhancement; at this point, incident photons will resonate with the scattering photon energy. Excited plasmon resonance radiates the scattering photons once again. The twofold EM enhancement theory estimates about 10 [14] enhancement of Raman scattering intensity of the incident light intensity. The equation for the enhancement of the Raman intensity is

$$I \propto \frac{8\pi(\omega \pm \omega_{IF})^4}{9c^4} I_L L^2(\omega) L^2(\omega_{IF}) \sum_{\rho,\sigma} |\alpha_{\rho\sigma}|^2 \qquad (\alpha_{\sigma,\rho} = x, y, z). \tag{1.11}$$

Here, L is EM enhancement factor. The EM enhancement factor as a function of the light energy shows an intense field at the incident (ω) and/or scattering (ω_{IF}) energy. In this theory, L is also concerned with the Raman enhancement in each polarisation direction. The polarisation direction is also important in determining both the Raman enhancement factor and the active mode of vibration.

1.3.3 Electronic (and Vibronic) Resonance Raman Scattering

When the incident light energy is close to a particular molecular absorption band, the intensity of Raman scattering shows enhancement in some orders as an incident photon field. In quantum mechanical theory, the polarisability tensor is derived from the Kramers–Heisenberg–Dirac (KHD) dispersion equation as follows:

$$\alpha_{\rho\sigma} = \sum_{e \neq m,n} \left\{ \frac{\langle I \mid \mu_\sigma | K \rangle \langle K \mid \mu_\rho | F \rangle}{E_K - E_I - \hbar\omega - i\Gamma} + \frac{\langle I \mid \mu_\rho | K \rangle \langle K \mid \mu_\sigma | F \rangle}{E_K - E_F + \hbar\omega + i\Gamma} \right\}, \tag{1.12}$$

where $|I\rangle$ is the ground state, $|F\rangle$ is the final state and $|K\rangle$ is the intermediate state. In the normal (nonresonant) Raman scattering, $E_K - E_I\rangle\rangle\hbar\omega$ is satisfied, and the term of the energy denominator $E_K - E_I - \hbar\omega$ become almost the same for all intermediate states. In resonance Raman scattering (RRS), $E_K - E_I \approx \hbar\omega$ is consisted, and the specific vibrational mode with a small energy denominator $E_K - E_I - \hbar\omega$ shows the Raman enhancement in some order.

In the nonresonant condition, it is sufficient to use the Born–Oppenheimer approximation. This approximation neglects coupling between electronic and nuclear motions. However, in the resonant condition, the electronic transition moment has a slight dependence on the normal coordinates of vibration Q, and Hamiltonian coupling may be expressed by an expansion in the nuclear displacement (Herz–Teller expansion). We obtain the polarisability tensor in the form

$$\alpha_{\rho\sigma} = A + B + C + D, \tag{1.13}$$

where

$$A = \sum_{K \neq I} \sum_{k} \frac{M_{KI}^{\sigma}(Q_0) M_{KI}^{\rho}(Q_0)}{\hbar(\omega_{KI} - \omega) - i\Gamma} \langle i \mid k \rangle \langle k \mid f \rangle, \tag{1.14}$$

$$B = \sum_{K\neq I}\sum_{k}\sum_{M\neq K}\left[\frac{M^{\sigma}_{IK}h_{KM}M^{\rho}_{MI}}{\hbar(\omega_{KI}-\omega)-i\Gamma}\right]\frac{\langle i\mid k\rangle\langle k\mid Q|f\rangle}{\hbar\omega_{MK}},$$
$$+\sum_{K\neq I}\sum_{k}\sum_{M\neq K}\left[\frac{M^{\sigma}_{IM}h_{MK}M^{\rho}_{KI}}{\hbar(\omega_{KI}-\omega)-i\Gamma}\right]\frac{\langle i\mid Q|k\rangle\langle k\mid f\rangle}{\hbar\omega_{MK}} \tag{1.15}$$

$$C = \sum_{K\neq I}\sum_{k}\sum_{M\neq I}\left[\frac{h_{IM}M^{\sigma}_{MK}M^{\rho}_{KI}}{\hbar(\omega_{KI}-\omega)-i\Gamma}\right]\frac{\langle i\mid k\rangle\langle k\mid Q|f\rangle}{\hbar\omega_{IM}},$$
$$+\sum_{K\neq I}\sum_{k}\sum_{M\neq I}\left[\frac{M^{\sigma}_{IK}M^{\rho}_{KM}h_{MI}}{\hbar(\omega_{KI}-\omega)-i\Gamma}\right]\frac{\langle i\mid Q|k\rangle\langle k\mid f\rangle}{\hbar\omega_{IM}} \tag{1.16}$$

$$D = \sum_{K\neq M}\sum_{k}\sum_{I\neq M}\left[\frac{M^{\sigma}_{IK}h_{KM}h_{MK}M^{\rho}_{KI}}{\hbar(\omega_{IM}-\omega)-i\Gamma}\right]\frac{\langle i\mid Q|k\rangle\langle k\mid Q|f\rangle}{\hbar^{2}\omega_{MK}}. \tag{1.17}$$

In Fig. 1.7, each transition electronic dipole for Raman processes is described. The *A*-term is described by the Franck–Condon factor $\langle i \mid k\rangle\langle k \mid f\rangle$. The *B*-term illustrates the degree of vibronic coupling between one electronic state and another. Each vibronic coupling is shown by the dotted line in Fig. 1.7. As the energy separation of the ground electronic state and the excited electronic state is normally large, the *C*- and *D*-terms are likely to be negligible, and we will not consider them in the resonant Raman condition.

1.3.4 Chemical Effect of Surface-enhanced Raman Scattering

In the SERS system, there is some additional enhancement in the nontotally symmetric mode of vibration. This appearance of an additional mode is not

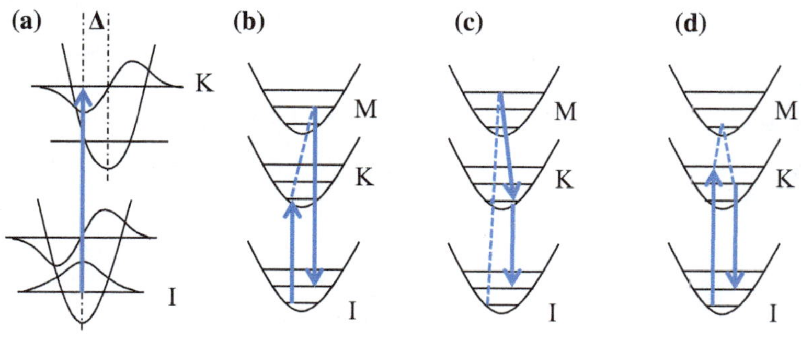

Fig. 1.7 The transition electronic dipole products that occur in the numerator of the selected terms of *A*, *B*, *C* and *D* in Eqs. (1.14)–(1.17)

explained by the mechanism of the EM effect. This origin is known to be a chemical effect via the charge transfer (CT) from metal (molecule) to molecule (metal). When the resonant condition is satisfied between the CT energy and excitation energy, polarisation is shown as

$$\alpha_{\rho\sigma} = A_f + A_k + B + C, \tag{1.18}$$

where

$$A_f = \frac{2}{\hbar}\sum_M M^{\sigma}_{MI} M^{\rho}_{MI} \langle i \mid k\rangle\langle k \mid f\rangle \frac{\omega_{MI} + \omega_f}{(\omega_{MI} + \omega_f)^2 - \omega^2}, \tag{1.19}$$

$$A_k = \frac{2}{\hbar}\sum_M M^{\sigma}_{KM} M^{\rho}_{KM} \langle i \mid k\rangle\langle k \mid f\rangle \frac{\omega_{KM} + \omega_k}{(\omega_{KM} + \omega_k)^2 - \omega^2}, \tag{1.20}$$

$$B = -\frac{2}{\hbar}\sum_{K\neq I}\sum_{M<K} M^{\sigma}_{KI} M^{\rho}_{MI} h_{KM} \langle i \mid Q|f\rangle \frac{(\omega_{KI}\omega_{MI} + \omega^2)}{(\omega^2_{KI} - \omega)(\omega^2_{MI} - \omega)}, \tag{1.21}$$

$$C = -\frac{2}{\hbar}\sum_{K\neq I}\sum_{M>I} M^{\sigma}_{MK} M^{\rho}_{KI} h_{IM} \langle i \mid Q|f\rangle \frac{(\omega_{KI}\omega_{KM} + \omega^2)}{(\omega^2_{KI} - \omega^2)(\omega^2_{KM} - \omega^2)}. \tag{1.22}$$

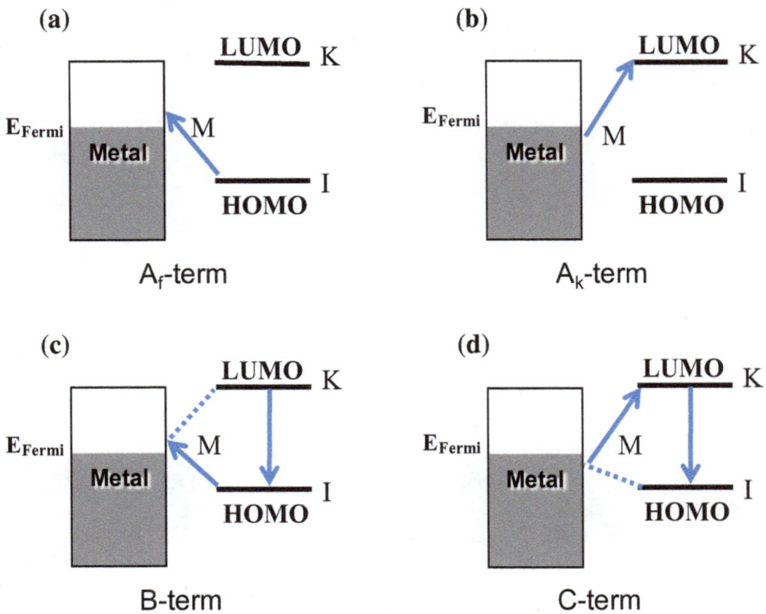

Fig. 1.8 Transition electronic dipole products that occur in the numerator for terms A_f, A_k, B and C in Eqs. (1.18)–(1.22)

These resonant conditions are shown schematically in Fig. 1.8. The transition electronic dipole is shown by the arrowed line. When the excitation energy is resonant to each transition energy, the term in the polarisability tensor is enhanced, and the Raman activity increases.

1.4 The Aim of the Study

If the molecules situated on the metal nanoparticle surface shine light, many photoexcitation processes will occur. Moreover, plasmon-enhanced phenomena can be used to easily obtain molecular vibrational information via Raman scattering.

First, I fabricate a highly ordered nanostructure and evaluated the polarisation of the SERS scattering photons (Chap. 2). Following this, conductance measurement at the junction of the metal–molecule–metal system proves that SERS depolarisation and CT behaviour occurred in the interface between the metal and molecules (Chap. 3). In the electrochemical environment, photoexcitation processes are evaluated using single-walled carbon nanotubes in the metal nanogap (Chap. 4). For more active modulation of the photoexcitation process in the hybrid molecule–metal system, a strong-coupling regime between molecular excitons and plasmon resonance of metal nanoparticles is achieved (Chap. 5). The resonant condition in the hybridised state is controlled using electrochemical potential tuning (Chap. 6).

The results of this study will offer a novel molecular photoexcitation process that is localised near the metal surface. Moreover, the new hybrid state of LSP is expected to modulate the intrinsic energy exciting the molecules or plasmonic metal nanoparticles. These techniques give the insight into the generation of a new photoexcitation energy state at the interface.

References

1. A. Otto, Z. Phys. **216**, 398 (1968)
2. E. Kretschman, H. Raether, Z. Naturforsch. **23a**, 2135 (1968)
3. E. Kretschmann, Z. Phys. **241**, 313 (1971)
4. R.H. Ritchie, E.T. Arakawa, J.J. Cowan, R.N. Hamm, Phys. Rev. Lett. **21**, 1530–1533 (1968)
5. I. Pockrand, Opt. Commun. **13**, 311–313 (1975)
6. T.W. Ebbesen, H.J. Lezec, H.F. Ghaemi, T. Thio, P.A. Wolff, Nature **391**, 667–669 (1998)
7. H.F. Ghaemi, T. Thio, D.E. Grupp, T.W. Ebbesen, H.J. Lezec, Phys. Rev. B **58**, 6779–6782 (1998)
8. T. Ishi, J. Fujikata, K. Ohashi, Jpn. J. Appl. Phys., Part 2 **44**, L170–L172 (2005)
9. I.R. Hooper, J.R. Sambles, Phys. Rev. B **70**, 045421 (2004)
10. D. Sarid, Phys. Rev. Lett. **47**, 1927–1930 (1981)
11. M. Fukui, V.C.Y. So, R. Normandin, Phys. Status Solidi (b) **91**, K61–K64 (1979)
12. N.J. Halas, S. Lal, W.-S. Chang, S. Link, P. Nordlander, Chem. Rev. **111**, 3913–3961 (2011)
13. N.C. Lindquist, J. Jose, S. Cherukulappurath, X. Chen, T.W. Johnson, S.-H. Oh, Laser Photonics Rev. **7**, 453–477 (2013)

14. D. Bartczak, O.L. Muskens, T. Sanchez-Elsner, A.G. Kanaras, T.M. Millar, ACS Nano **7**, 5628–5636 (2013)
15. V.V. Terekhin, A.V. Zaitseva, O.V. Dement'eva, V.M. Rudoy, Colloid J. **75**, 437–443 (2013)
16. N.J. Borys, E. Shafran, J.M. Lupton, Sci. Rep. **3** (2013)
17. L. Le Thi Ngoc, M. Jin, J. Wiedemair, A. van den Berg, E.T. Carlen, ACS Nano **7**, 5223–5234 (2013)
18. H. Liu, Y.M. Liu, T. Li, S.M. Wang, S.N. Zhu, X. Zhang, Phys. Status Solidi (b) **246**, 1397–1406 (2009)
19. N. Liu, H. Giessen, Angew. Chem. Int. Ed. **49**, 9838–9852 (2010)
20. C. Xue, Y. Xue, L. Dai, A. Urbas, Q. Li, Adv. Opt. Mat. **1**, 581–587 (2013)
21. R.W. Wood, Phil. Mag. **4**, 396 (1902)
22. J.A. Fan, C. Wu, K. Bao, J. Bao, R. Bardhan, N.J. Halas, V.N. Manoharan, P. Nordlander, G. Shvets, F. Capasso, Science **328**, 1135–1138 (2010)
23. S. Zhang, D.A. Genov, Y. Wang, M. Liu, X. Zhang, Phys. Rev. Lett. **101**, 047401 (2008)
24. S.J. Smith, E.M. Purcell, Phys. Rev. **92**, 1069–1069 (1953)
25. G.L. Liu, Y.-T. Long, Y. Choi, T. Kang, L.P. Lee, Nat. Methods **4**, 1015–1017 (2007)
26. J.R. Lakowicz, K. Ray, M. Chowdhury, H. Szmacinski, Y. Fu, J. Zhang, K. Nowaczyk, Analyst **133**, 1308–1346 (2008)
27. T. Ming, L. Zhao, H. Chen, K.C. Woo, J. Wang, H.-Q. Lin, Nano Lett. **11**, 2296–2303 (2011)
28. S. Hayashi, K. Kozaru, K. Yamamoto, Solid State Commun. **79**, 763–767 (1991)
29. H. Shinji, O. Takayuki, J. Phys. D Appl. Phys. **45**, 433001 (2012)
30. H.R. Stuart, D.G. Hall, Appl. Phys. Lett. **69**, 2327–2329 (1996)
31. H.R. Stuart, D.G. Hall, Appl. Phys. Lett. **73**, 3815–3817 (1998)
32. T. Neal, K. Okamoto, A. Scherer, Opt. Express **13**, 5522–5527 (2005)
33. P. Anger, P. Bharadwaj, L. Novotny, Phys. Rev. Lett. **96**, 113002 (2006)
34. K.R. Catchpole, S. Pillai, J. Lumin. **121**, 315–318 (2006)
35. S. Pillai, K.R. Catchpole, T. Trupke, M.A. Green, J. Appl. Phys. **101**, 093105–093105–093105–093108 (2007)
36. D. Derkacs, S.H. Lim, P. Matheu, W. Mar, E.T. Yu, Appl. Phys. Lett. **89**, 093103 (2006)
37. A. Köck, W. Beinstingl, K. Berthold, E. Gornik, Appl. Phys. Lett. **52**, 1164–1166 (1988)
38. A. Köck, E. Gornik, M. Hauser, W. Beinstingl, Appl. Phys. Lett. **57**, 2327–2329 (1990)
39. N.E. Hecker, R.A. Höpfel, N. Sawaki, T. Maier, G. Strasser, Appl. Phys. Lett. **75**, 1577–1579 (1999)
40. G.A. Plotz, H.J. Simon, J.M. Tucciarone, J. Opt. Soc. Am. **69**, 419–422 (1979)
41. J. Seidel, S. Grafström, L. Eng, Phys. Rev. Lett. **94**, 177401 (2005)
42. M.A. Noginov, V.A. Podolskiy, G. Zhu, M. Mayy, M. Bahoura, J.A. Adegoke, B.A. Ritzo, K. Reynolds, Opt. Express **16**, 1385–1392 (2008)
43. M.A. Noginov, G. Zhu, M. Mayy, B.A. Ritzo, N. Noginova, V.A. Podolskiy, Phys. Rev. Lett. **101**, 226806 (2008)
44. J. Hofmann, W. Steinmann, Phys. Status Solidi (b) **30**, K53–K56 (1968)
45. T. Inagaki, K. Kagami, E.T. Arakawa, Phys. Rev. B **24**, 3644–3646 (1981)
46. T. Inagaki, K. Kagami, E.T. Arakawa, Appl. Opt. **21**, 949–954 (1982)
47. J. Lehmann, M. Merschdorf, W. Pfeiffer, A. Thon, S. Voll, G. Gerber, Phys. Rev. Lett. **85**, 2921–2924 (2000)
48. V.M. Shalaev, C. Douketis, J.T. Stuckless, M. Moskovits, Phys. Rev. B **53**, 11388–11402 (1996)
49. J.G. Endriz, W.E. Spicer, Phys. Rev. Lett. **24**, 64–68 (1970)
50. P.L. Redmond, L.E. Brus, J. Phys. Chem. C **111**, 14849–14854 (2007)
51. R. Jin, Y. Cao, C.A. Mirkin, K.L. Kelly, G.C. Schatz, J.G. Zheng, Science **294**, 1901–1903 (2001)
52. R. Jin, Y.C. Cao, E. Hao, G.S. Metraux, G.C. Schatz, C.A. Mirkin, Nature **425**, 487–490 (2003)
53. L. Brus, Acc. Chem. Res. **41**, 1742–1749 (2008)

54. X. Wu, E.S. Thrall, H. Liu, M. Steigerwald, L. Brus, J. Phys. Chem. C **114**, 12896–12899 (2010)
55. Y. Tian, T. Tatsuma, Chem. Commun. 1810–1811 (2004)
56. Y. Tian, T. Tatsuma, J. Am. Chem. Soc. **127**, 7632–7637 (2005)
57. Y. Nishijima, K. Ueno, Y. Yokota, K. Murakoshi, H. Misawa, J. Phys. Chem. Lett. **1**, 2031–2036 (2010)
58. Y. Nishijima, K. Ueno, Y. Kotake, K. Murakoshi, H. Inoue, H. Misawa, J. Phys. Chem. Lett. **3**, 1248–1252 (2012)
59. I. Pockrand, A. Brillante, D. Mobius, J. Chem. Phys. **77**, 6289–6295 (1982)
60. D.G. Lidzey, D.D.C. Bradley, M.S. Skolnick, T. Virgili, S. Walker, D.M. Whittaker, Nature **395**, 53–55 (1998)
61. D.G. Lidzey, Science **288**, 1620–1623 (2000)
62. J. Bellessa, C. Bonnand, J. Plenet, J. Mugnier, Phys. Rev. Lett. **93**, 036404 (2004)
63. J. Dintinger, S. Klein, F. Bustos, W. Barnes, T. Ebbesen, Phys. Rev. B **71**, 035424 (2005)
64. Y. Sugawara, T.A. Kelf, J.J. Baumberg, M.E. Abdelsalam, P.N. Bartlett, Phys. Rev. Lett. **97**, 266808 (2006)
65. T. Hakala, J. Toppari, A. Kuzyk, M. Pettersson, H. Tikkanen, H. Kunttu, P. Törmä, Phys. Rev. Lett. **103**, 053602 (2009)
66. J.A. Hutchison, T. Schwartz, C. Genet, E. Devaux, T.W. Ebbesen, Angew. Chem. Int. Ed. **51**, 1592–1596 (2012)
67. J.A. Hutchison, A. Liscio, T. Schwartz, A. Canaguier-Durand, C. Genet, V. Palermo, P. Samori, T.W. Ebbesen, Adv. Mater. **25**, 2481–2485 (2013)
68. G.P. Wiederrecht, G.A. Wurtz, J. Hranisavljevic, Nano Lett. **4**, 2121–2125 (2004)
69. G.A. Wurtz, P.R. Evans, W. Hendren, R. Atkinson, W. Dickson, R.J. Pollard, A.V. Zayats, W. Harrison, C. Bower, Nano Lett. **7**, 1297–1303 (2007)
70. N.T. Fofang, T.H. Park, O. Neumann, N.A. Mirin, P. Nordlander, N.J. Halas, Nano Lett. **8**, 3481–3487 (2008)
71. W. Ni, T. Ambjornsson, S.P. Apell, H. Chen, J. Wang, Nano Lett. **10**, 77–84 (2010)
72. A.E. Schlather, N. Large, A.S. Urban, P. Nordlander, N.J. Halas, Nano Lett. **13**, 3281–3286 (2013)
73. M. Inoue, K. Ohtaka, J. Phys. Soc. Jpn. **52**, 3853–3864 (1983)
74. B. Pettinger, J. Chem. Phys. **85**, 7442–7451 (1986)
75. H. Xu, E.J. Bjerneld, M. Käll, L. Börjesson, Phys. Rev. Lett. **83**, 4357–4360 (1999)
76. H. Xu, X.-H. Wang, M.P. Persson, H.Q. Xu, M. Käll, P. Johansson, Phys. Rev. Lett. **93**, 243002 (2004)
77. A.L. Falk, F.H.L. Koppens, C.L. Yu, K. Kang, N. de Leon Snapp, A.V. Akimov, M.-H. Jo, M.D. Lukin, H. Park, Nat. Phys. **5**, 475–479 (2009)

Chapter 2
The Depolarisation Behaviour of Surface-Enhanced Raman Scattering Photons in a Metal Nanodimer Structure

2.1 Introduction

The metal–molecule interface provides highly attractive fields that interact with each other, such as in charge transfer. However, little is known about the materials localised at the interface. Polarised Raman measurement represents an attractive research field related to metal surfaces. The polarised SERS signal has a highly polarised character because of the anisotropic EM feature on the metal surface. Moreover, such signals are considered to have high sensitivity to the electronic state at the interface. Polarised SERS measurements give information about both the incident and Raman scattering polarisations, and several interesting polarised SERS studies have been carried out [1–13]. If one assumes a symmetrical single nanodimer structure in which only the localised EM field at the gap contributes to the enhancement, SERS depolarisation is not observed because the polarisations for excitation and scattering are the same.

Linearly polarised scattered light was observed in a metal nanodimer system by Haran et al., who reported that an asymmetric metal nanoparticle trimer generates elliptically polarised scattered light; thus, these researchers showed that the polarisation of light scattered from molecules can be manipulated on the nanometre scale [7]. A well-accepted model of the twofold EM enhancement theory for excitation and scattering for the SERS process cannot explain the present observations, because the theory predicts that the polarisation of the scattering photons is controlled in the same manner as that of the localised field [3, 14]. In contrast, depolarisation behaviour has been also reported. Polarisation rotation induced by the plasmon resonance of the SERS signal with respect to the incident polarisation was shown in arrays of uniform Au nanostructures [11]. By using an elliptic nanostructure that exhibits two LSP resonances each associated with one of the two principle axes, it was found that scattered light is resonant with another axis to one axis of the excitation process.

F. Nagasawa, *Studies on the Plasmon-Induced Photoexcitation Processes of Molecules on Metal Surfaces*, Springer Theses,
DOI 10.1007/978-4-431-56579-6_2

In the present study, the polarisation dependence on SERS from a well-ordered Ag dimer array [15] was measured in an aqueous solution containing the target molecule, 4,4'-bipyridine. In this chapter, the polarisation dependence of the scattering photons is discussed with respect to the optical properties of the metal nanostructure and adsorption structure of the target molecule.

2.2 Experimental

The Ag dimer arrays were prepared via an angle-resolved nanosphere lithography (AR-NSL) technique using repeated vapour depositions onto a polystyrene (PS)-particle (Polysciences Inc.; diameter = 350 nm) monolayer prepared on a glass substrate [16]. Aqueous PS particle suspension was concentrated to 10 wt% and then diluted by ethanol (50%). As well as the previously documented drop-coating method, I adapted another method using PS monolayers prepared on liquid–gas interfaces to improve the quality of the array. The PS solutions (20 μL) were dropped on the convex surface of a watch glass immersed in Milli-Q water. The thin water layer on the glass surface led to uniform spreading of PS beads on the air–water interface. After the spreading of PS beads on the water surface, the layer was packed tight in the course of nature. The change in the surface tension of the water around the PS layer resulted in the formation of well-ordered monolayers with a low density of defects, dislocations and vacancies on the liquid–water interface. The prepared high-quality monolayer was then lifted off the water surface using a clean glass substrate. The metal was deposited onto the PS monolayer prepared on the glass substrate. After the first metal deposition, the second metal was deposited at different angles. Then, the PS mask was removed by sonication in Milli-Q water for 10–30 s. The extinction spectrum of the metal dimer array in the visible-near infrared (NIR) region was recorded utilising a multichannel spectrometer (MCPD-2000, Ohtsuka Electronics). The structure of the dimer on the glass substrate was inspected using an atomic force microscope (AFM, Nanoscope-IIIa, Digital Instruments) in air.

A homemade polarised Raman microprobe spectrometer was specially modified for NIR laser light (λ_{ex} = 785 nm). Raman measurements were carried out using the backscattering configuration by simultaneously collecting the scattering photons with parallel and perpendicular polarisation directions (Fig. 2.1a). The expanded NIR beam was focussed onto the sample using a water-immersion objective lens with 100× magnification and a numerical aperture of 1.0. The estimated spot size of irradiation of 1 μm with tunable output intensity was in the range of 10 μW–20 mW. All Raman measurements were carried out in situ via immersion of the SERS active substrates into aqueous solutions containing target molecules (4,4'-bipyridine; reagent grade, Wako Co. Ltd.) with controlled concentrations (1 μM–1 mM). The measurement on a single crystal was carried out in air by controlling the excitation polarisation direction to the crystal axis determined by XRD.

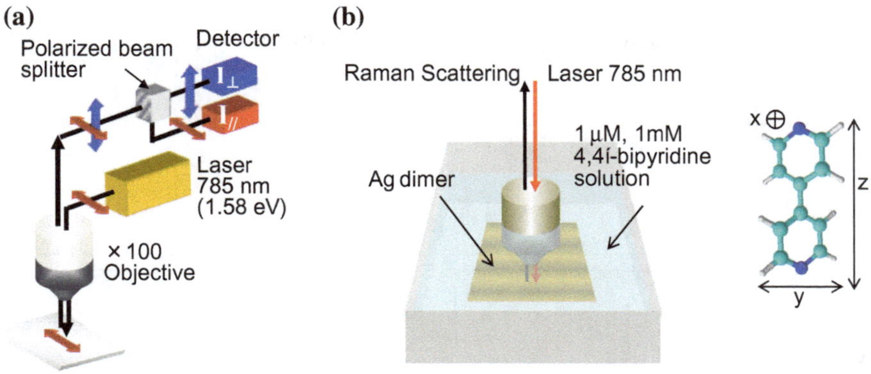

Fig. 2.1 Raman measurement configuration **a** in situ Raman cell and **b** 4,4'-bipyridine molecular structure

The stability of the metal structure was verified through polarised Rayleigh scattering light measurement. The identical polarised properties were obtained before and after light irradiation, indicating that the metal dimer structure is stable regardless of the light irradiation.

Density functional theory (DFT) calculations were carried out using the Gaussian 03, revision E.01, at the B3LYP level of DFT with 6-31G** basis sets. Calculations were carried out without the contribution of the resonance of 4,4'-bipyridine molecules. The Raman intensity was estimated as follows:

$$I \propto \nu_s^4 (e_s \alpha e_i)^2, \tag{2.1}$$

where ν_s is the wavenumber of the Raman scattering, e_s and e_i are the unit vector along the scattering and incident electric field and α is the polarisation tensor.

2.3 Results and Discussion

For Raman band assignments, conventional polarised Raman measurements were carried out using a 15 mM aqueous solution of 4,4'-bipyridine and tetrachloromethane solution (I_{ex} = 20 mW, t_{ex} = 100 s). The Raman spectrum of an aqueous solution for the parallel scattering configuration showed intense peaks at 1010 and 1298 cm^{-1} and weak peaks at 764, 873 and 1229 cm^{-1} (Fig. 2.2a). For the perpendicular configuration, scattering was not clearly observed (Fig. 2.2b). This indicates that the peaks assigned to the totally symmetric modes of 4,4'-bipyridine molecules [15, 17–19] had a zero depolarisation ratio for aqueous solution. The polarised Raman spectra of tetrachloromethane solution are shown in Fig. 2.2c, d. The Raman spectrum of the tetrachloromethane solution for the parallel scattering configuration showed intense peaks at 218 (E), 314 (T_2) and

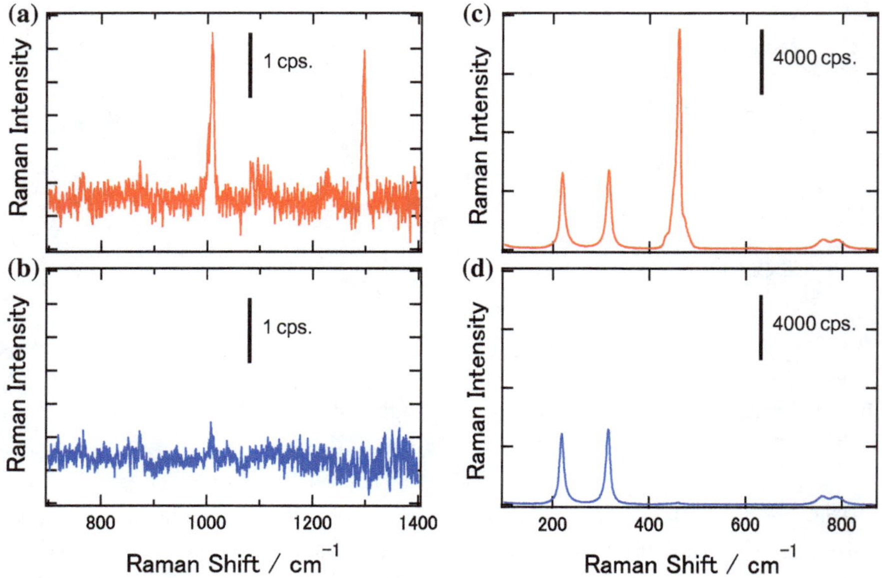

Fig. 2.2 **a**, **b** Polarised Raman spectrum in 4,4'-bipyridine solution and **c**, **d** tetrachloromethane solution: **a**, **c** parallel configuration and **b**, **d** perpendicular polarisation

460 cm^{-1} (A). Parallel scattering polarisation, which has a three-fourth depolarisation ratio for nontotally symmetric modes [218 (E), 314 cm^{-1} (T_2)] with linearly polarised incident light were measured.

To investigate the Raman polarised tensor, I fabricated a single 4,4'-bipyridine crystal (Fig. 2.3a). XRD analyses showed the P_{21} symmetry in Fig. 2.3b, which is a similar symmetry to that reported previously [20]. I measured each polarisation configuration to the single crystal axis of 4,4'-bipyridine and compared this to the DFT calculation. Figure 2.3c shows the polarised Raman spectra of each polarisation configuration (I_{ex} = 20 mW, t_{ex} = 10 s). For parallel polarisation configuration on a-excitation (excitation polarisation parallel to the a-axis of a single crystal) and a-scattering (scattering polarisation parallel to the a-axis) direction, relatively strong peaks at 762, 1000 and 1301 cm^{-1} and weak peaks at 880, 1075 and 1230 cm^{-1} were observed. For the other parallel configurations on (b, b) and (c, c), the same peaks were observed and the intensity ratios of the peaks were changed. The (b, b) configuration had the maximum peak intensity at 1240 cm^{-1}. In contrast, the (c, c) configuration had the maximum peak intensity at 1000 and 1300 cm^{-1}. For the perpendicular polarisation configuration on (a, b), additional peaks at 862, 1042, 1215 and 1325 cm^{-1} were observed. The polarisation dependence of the intensity ratio for the parallel polarisation and appearance of additional peaks for the perpendicular polarisation were well reflected in the calculated polarised Raman spectra (Fig. 2.3d). The perpendicular configurations were

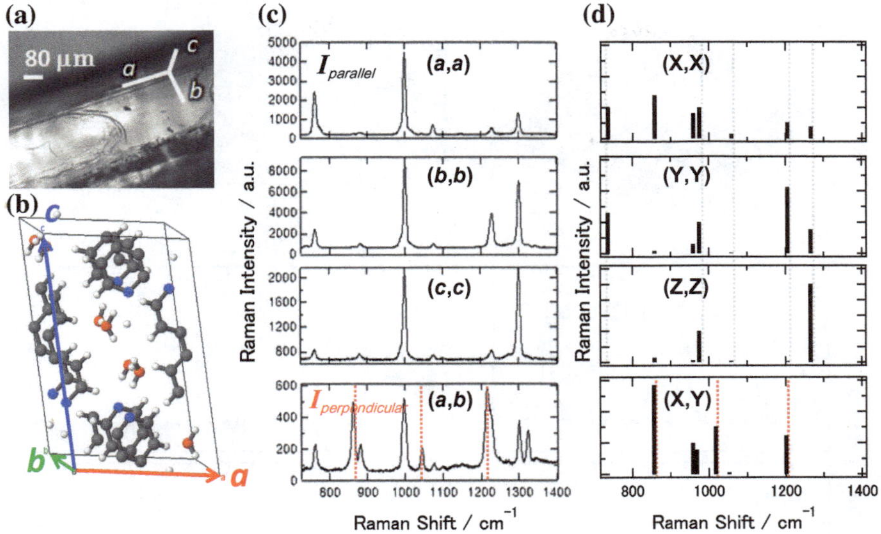

Fig. 2.3 **a** Photograph of a single 4,4'-bipyridine crystal, **b** crystal axis of the single 4,4'-bipyridine crystal, **c** polarised Raman spectrum of the single crystal in the parallel configuration and perpendicular configuration. The incident and scattering polarisation axes are shown with the notation (incident polarisation, scattering polarisation). **d** Calculated polarised Raman spectrum for parallel and perpendicular polarisation

assigned to totally symmetric *a* modes and the nontotally symmetric b1 modes of 4,4'-bipyridine molecules, respectively [15, 17–19].

The polarisation dependence on SERS was measured using the substrate with an Ag dimer array (Fig. 2.4a). The thickness and distance of the dimer structure were optimised to provide an extinction peak at around 780 nm at the parallel polarisation along the long dimer axis (Fig. 2.4b) for the excitation of the Raman measurement (I_{ex} = 785 nm). The dimer structure consisted of the different size Ag triangle, with a size is 50, 100 nm at each side of triangle and a height of 25, 30 nm. The gap sizes were estimated through comparison of the calculation [21]. From the extinction spectra at Fig. 2.4b, it can be observed that the polarised excitation along the long axis led to the appearance of a peak at a longer wavelength (around 800 nm). This agrees well with the theoretically calculated spectra of 'tip-to-side' Ag triangular dimers with distances of a few nanometres. Thus, the AR-NSL dimers had a gap distance of a few nanometres. The spot size irradiated to the substrate was estimated at ~1 μm, and this area included about 9–10 dimers. Immersed in aqueous solution containing 1 mM 4,4'-bipyridine, this substrate showed an intense SERS [15]. Polarised SERS measurements on the Ag dimmers at excitation parallel to the dimer axis exhibited well-defined polarisation behaviour (Fig. 2.4c, d). The SERS spectrum at the parallel polarisation configuration with excitation along the long dimer axis showed intense SERS peaks at 770, 871, 1016, 1074, 1231 and 1296 cm^{-1} (Fig. 2.4c). The observed SERS peaks at the parallel polarisation

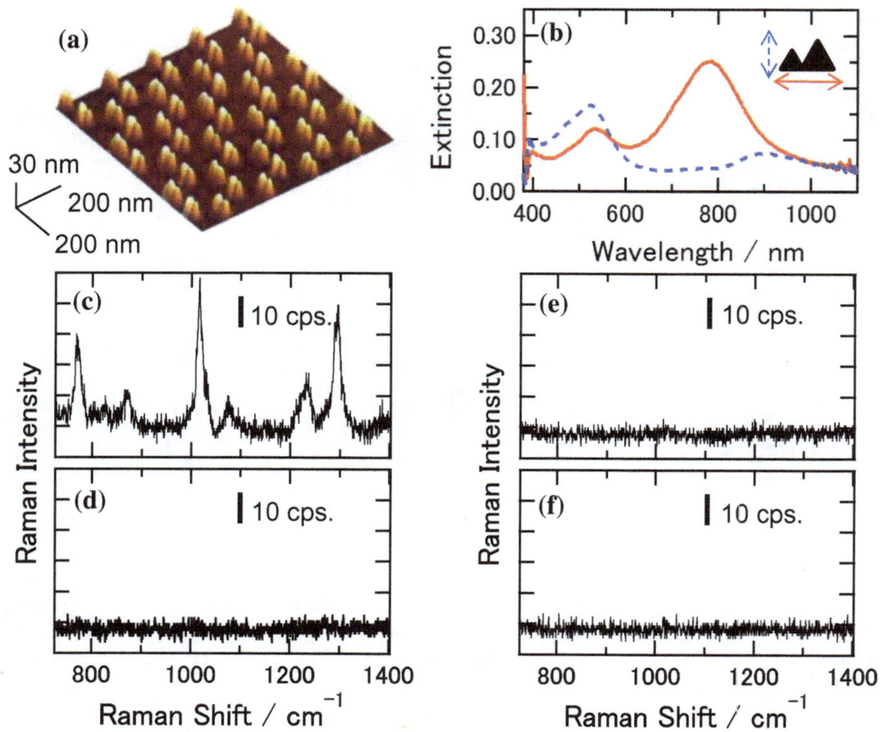

Fig. 2.4 **a** Atomic force microscope (AFM) image, **b** polarised extinction spectrum of Ag nanodimer array and SERS spectra observed in a 1 mM 4,4'-bipyridine aqueous solution at **c** parallel and **d** perpendicular polarisation configurations with parallel excitation to the dimer axis and **e** parallel and **f** perpendicular configurations with perpendicular excitation to the dimer axis (I_{ex} = 10 mM, t_{ex} = 1 s)

configuration could be assigned to the totally symmetric *a* modes of 4,4'-bipyridine. These peaks were not observed for the measurement on the scattering at the perpendicular configuration to the long dimer axis (Fig. 2.4d). At the perpendicular excitation to the dimer axis, scattering was not observed in either the parallel or the perpendicular configuration (Fig. 2.4e, f).

This apparent difference between the polarisation directions at the excitation demonstrated the contribution of the localised EM field at the gap of the dimers. The present system showed unique behaviour in which the polarisation of the scattering was dependent on the SERS active site and the concentration of the solution. Figure 2.5 depicts polarised SERS spectra observed in 1 mM and 1 mM 4,4'-bipyridine solutions at distinct SERS active sites under excitation parallel to the long dimer axis. At site A, perpendicular polarised scattering was not observed, as in the case shown in Fig. 2.4. Increasing the concentration to 1 mM at the same site still resulted in no perpendicular scattering (Fig. 2.5b). At site B, relatively weak perpendicular scattering was observed both in 1 μM and 1 mM solutions (Fig. 2.5c, d).

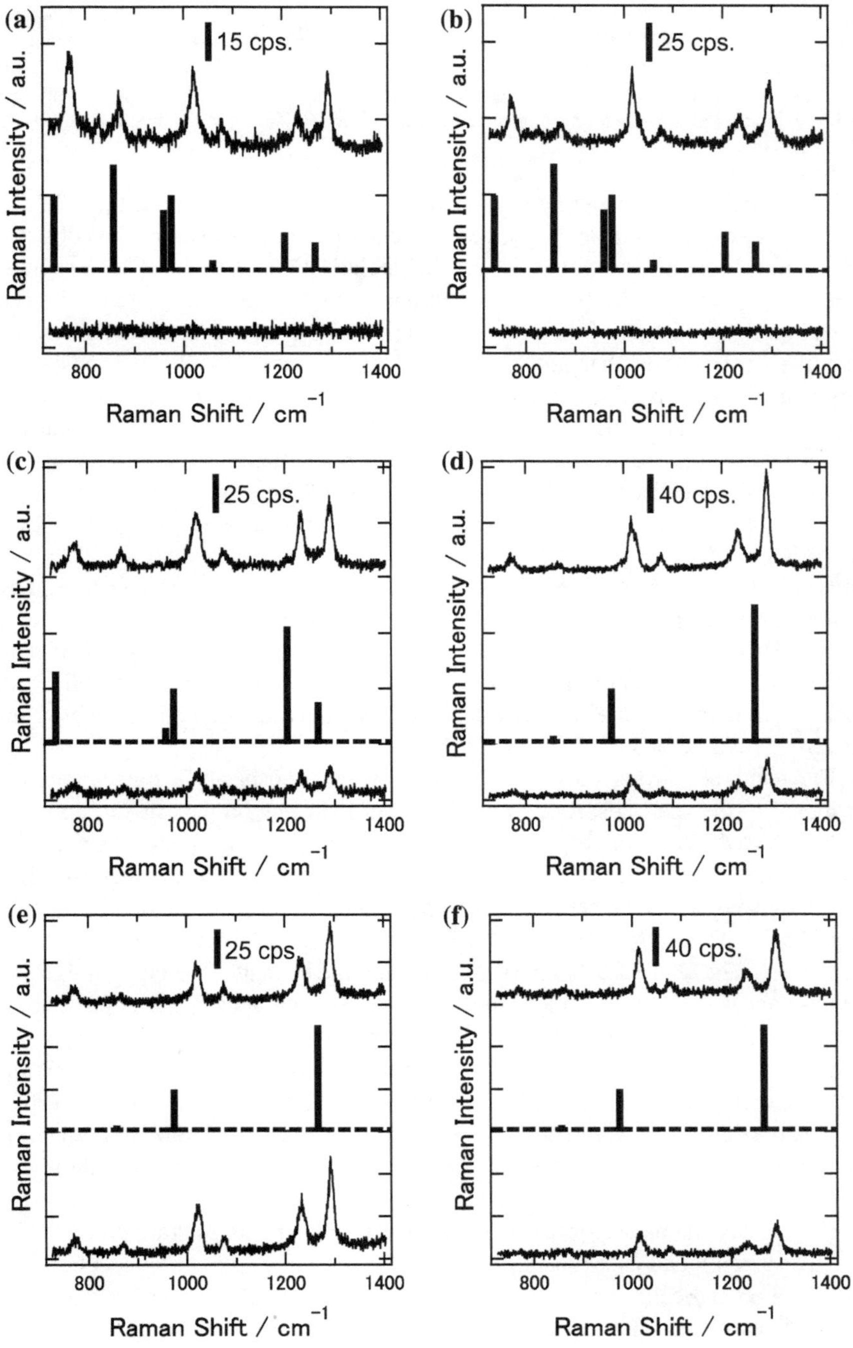
(a)
15 cps.
(b)
25 cps.
(c)
25 cps.
(d)
40 cps.
(e)
25 cps.
(f)
40 cps.
Raman Intensity / a.u.
800
1000
1200
1400
Raman Shift / cm⁻¹

◀**Fig. 2.5** SERS spectra observed in **a**, **c**, **e** 1 mM and **b**, **d**, **f** 1 mM 4,4'-bipyridine aqueous solutions observed at **a**, **b** site A, **c**, **d** site B and **e**, **f** site C in the parallel (*upper*) and perpendicular (*lower*) polarisation configurations with excitation parallel to the dimer axis (I_{ex} = 10 mW, t_{ex} = 1 s); DFT calculation (*middle*) for polarisation of **a**, **b** X-excitation and X-scattering, **c**, **d** Y-excitation and Y-scattering and **d**, **e**, **f** Z-excitation and Z-scattering directions

The perpendicular scattering became much more apparent at site C (Fig. 2.5e, f). For the 1 mM solution (Fig. 2.5e), perpendicular polarised scattering was observed with comparable intensity to the parallel polarisation. For the 1 mM solution (Fig. 2.5f), perpendicular polarised scattering was weak, as in the case of site B. It should be noted that all of the observations in Fig. 2.5 were carried out before and after the confirmation that no SERS signal was observed with the excitation of perpendicular polarisation to the long dimer axis at the same site. Spectral features were also dependent on the orientation. Differences in the relative intensities between the bands and the wavenumbers should reflect the orientation of molecules relative to the localised EM field at the gap part of the dimer. Anisotropy of the highly localised EM field at the gap leads to a sensitive response in the intensity of the SERS band [22–24].

To discuss the orientation of molecules on the surface, spectral features were compared with those predicted by DFT calculation. SERS spectra at site A exhibited relatively intense bands at 770 and 871 cm^{-1} assigned to the out-of-plane ring and C–H deformation modes, respectively. These features were reproduced in DFT calculations for the polarisation in the X-direction for both excitation and scattering (middle of Fig. 2.5a, b).

The relatively strong out-of-plane modes versus the in-plane modes could be attributed to the 'flat' adsorption of 4,4'-bipyridine rings on the Ag surface.

A localised EM field perpendicular to the bipyridine rings on the surface may enhance the relative intensity of the band. At site B, the SERS spectrum observed at 1 mM showed a sharp increase in the relative intensity of the band at 1240 cm^{-1} (Fig. 2.5c). At 1 mM, evolution of the band at 1296 cm^{-1} was observed (Fig. 2.5d). The DFT calculations suggested that the band at 1240 cm^{-1} assigned to the in-plane C–H bending mode becomes intense with the 'flat' adsorption, in which the Y-axis of the molecule is normal to the metal surface. An intense band observed at 1300 cm^{-1} assigned to the interring stretching mode could be attributed to the 'standing' adsorption in which the molecular Z-axis is normal to the surface. The features at site C were comparable to those of site B in 1 mM solutions. The present band analysis proves the site dependence of the molecular orientation in the present system.

The relative intensities of the perpendicular to the parallel scattering in site C of Fig. 2.5 were plotted as a function of the 4,4'-bipyridine solution concentration in Fig. 2.6. Perpendicular scattering was observed, that is highly depolarised behaviour on SERS, especially at a low concentration. Depolarisation was observed when the spectral features corresponded to the 'flat' adsorption. To evaluate the orientation effect, we plotted the relative intensity against the Raman intensity ratios of $I_{780\,\mathrm{cm}^{-1}}/I_{1000\,\mathrm{cm}^{-1}}$ and $I_{1300\,\mathrm{cm}^{-1}}/I_{1000\,\mathrm{cm}^{-1}}$ in Fig. 2.7. As shown in the DFT

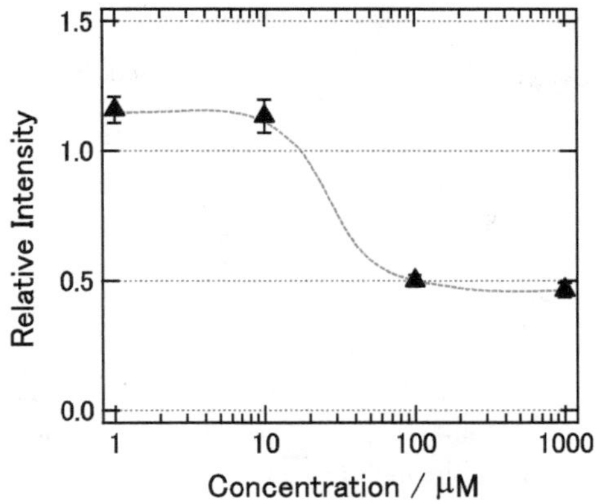

Fig. 2.6 Relative SERS intensity ratio of the perpendicular and parallel polarisation configurations with parallel excitation to the dimer axis at various 4,4'-bipyridine concentrations (site C)

calculation, the 'flat' adsorption of 4,4'-bipyridine rings on the Ag surface indicated an intense band at 780 cm^{-1}, and the 'standing' adsorption—in which the Y-axis of the molecule is normal to the metal surface—indicated an intense band at 1300 cm^{-1}. These correlations illustrate that the intense $I_{780\,cm^{-1}}/I_{1000\,cm^{-1}}$ corresponds to the 'flat' adsorption, and the intense $I_{1300\,cm^{-1}}/I_{1000\,cm^{-1}}$ corresponds to the 'standing' adsorption. 'Flat' adsorption tends towards low relative intensity in Fig. 2.7a. In contrast, 'standing' adsorption indicates high relative intensity in Fig. 2.7b. The present observation of the depolarisation behaviour clearly demonstrated that the orientation of molecules at the gap affected the polarisation of the scattering process. The polarised dependence of SERS has been discussed with

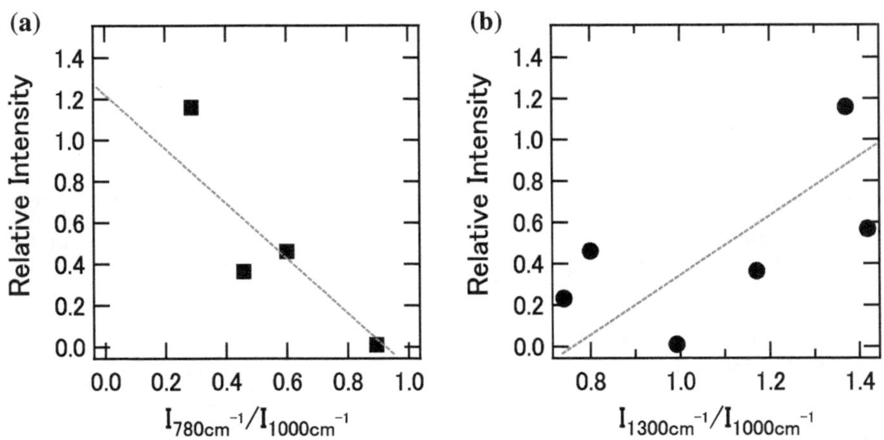

Fig. 2.7 Relative intensity ratio of the parallel and perpendicular polarisation configurations with parallel excitation to the dimer axis at Raman intensity ratios of $I_{780\,cm^{-1}}/I_{1000\,cm^{-1}}$ and $I_{1300\,cm^{-1}}/I_{1000\,cm^{-1}}$; all concentrations are 1 μM

respect to the individual tensors for the excitation field and the scattering field [7, 11]. To explain the present observation concerning the depolarisation of the scattering photons, selective contributions to the scattering process should be considered. One of the possible contributions is the two different resonance modes of LSPR, which respectively relate to incident and scattering polarisation. The extinction of the Ag dimer at a longer wavelength region >850 nm in perpendicular polarisation contributes to the resonance of the Raman scattering at perpendicular polarisation around 1300 cm^{-1} (874 nm). This perpendicular polarisation, as well as the nondiagonal terms in the polarisability tensor that depend on the orientation of adsorbed molecules, could induce the perpendicular scattering. This process, however, is not adequate to explain the present observation because of relatively small intensities of the extinction of the metal structure and the nontotally symmetric bands of a single crystal. Another possibility is that the scattering depolarisation observed in the present system could be induced by the resonance of the scattering photons with the localised electronic states caused by the specific adsorption. In the present system, 4,4'-bipyridine molecules adsorbed on the metal surface tended to have the 'standing' structure when the concentration was more than a few tens of mM due to the coordination ability of the nitrogen atom and the intermolecular interaction [25, 26]. Specific adsorption of molecules could lead to the formation of localised electronic states at the molecule and metal interface. The contribution of the localised resonance states to the SERS process has been considered as an excitation process involving the CT contribution, the so-called chemical effect, leading to the observation of nontotally symmetric modes [26]. Although the observed bands in the SERS spectra in the present system were fully assigned to totally symmetric modes, characteristics of the SERS spectral features, such as relatively intense bands at 1229 and 1298 cm^{-1} compared with those in bulk solution and a single crystal, could imply a CT contribution [17]. To demonstrate the contribution of the resonance of the localised states to the depolarisation, further detailed analysis is required using a system showing well-controlled single-molecule SERS.

2.4 Conclusion

In conclusion, the substrate with a Ag nanodimer array immersed in 4,4'-bipyridine solution showed depolarised SERS behaviour depending on the adsorbed structure of molecules at the dimer's gaps. The orientation of molecules was discussed by comparing SERS spectra with those obtained from conventional polarised Raman spectra of a homogeneous aqueous solution and a single crystal of 4,4'-bipyridine. The SERS depolarisation at specific adsorption states of molecules implies that the molecule and metal interface modulate the scattering polarisation selectively.

References

1. M. Moskovits, J.S. Suh, J. Phys. Chem. **88**, 1293–1298 (1984)
2. H. Sano, G. Mizutani, S. Ushioda, Phys. Rev. B **47**, 13773–13781 (1993)
3. K.A. Bosnick, Jiang, L.E. Brus, J. Phys. Chem. B **106**, 8096–8099 (2002)
4. Jiang, K. Bosnick, M. Maillard, L. Brus, J. Phys. Chem. B **107**, 9964–9972 (2003)
5. H. Xu, M. Kall, ChemPhysChem **4**, 1001–1005 (2003)
6. T.O. Shegai, G. Haran, J. Phys. Chem. B **110**, 2459–2461 (2006)
7. T. Shegai, Z. Li, T. Dadosh, Z. Zhang, H. Xu, G. Haran, Proc. Natl. Acad. Sci. U.S.A. **105**, 16448–16453 (2008)
8. Z. Li, T. Shegai, G. Haran, H. Xu, ACS Nano **3**, 637–642 (2009)
9. E.C. Le Ru, P.G. Etchegoin, Chem. Phys. Lett. **423**, 63–66 (2006)
10. E.C. Le Ru, M. Meyer, E. Blackie, P.G. Etchegoin, J. Raman Spectrosc. **39**, 1127–1134 (2008)
11. E.C. Le Ru, J. Grand, N. Félidj, J. Aubard, G. Lévi, A. Hohenau, J.R. Krenn, E. Blackie, P.G. Etchegoin, J. Phys. Chem. C **112**, 8117–8121 (2008)
12. A. Moroz, J. Opt. Soc. Am. B **26**, 517–527 (2009)
13. K.D. Jernshøj, S. Hassing, J. Raman Spectrosc. **41**, 727–738 (2010)
14. K.-I. Yoshida, T. Itoh, V. Biju, M. Ishikawa, Y. Ozaki, Phys. Rev. B **79**, 085419 (2009)
15. Y. Sawai, B. Takimoto, H. Nabika, K. Ajito, K. Murakoshi, J. Am. Chem. Soc. **129**, 1658–1662 (2007)
16. U.C. Fischer, H.P. Zingsheim, J. Vac. Sci. Tech. **19**, 881–885 (1981)
17. J.K. Lim, S.-W. Joo, Surf. Interface Anal. **39**, 684–690 (2007)
18. A. Topaçli, S. Akyüz, Spectrochim. Acta Part A Mol. Biomol. Spectrosc. **51**, 633–641 (1995)
19. A.L. Kamyshny, V.N. Zakharov, Y.V. Fedorov, A.E. Galashin, L.A.J. Aslanov, Colloid Interface Sci. **158**, 171–182 (1993)
20. C. Näther, J. Riedel, I. Jeß, Acta Crystallogr. Sect. C **57**, 111–112 (2001)
21. E. Hao, G.C. Schatz, J. Chem. Phys. **120**, 357–366 (2004)
22. D. Battisti, R. Aroca, R.O. Loutfy, Chem. Mat. **1**, 124–128 (1989)
23. P.G. Gucciardi, F. Bonaccorso, M. Lopes, L. Billot, M.L. de la Chapelle, Thin Solid Films **516**, 8064–8072 (2008)
24. K. Uosaki, H. Allen, O. Hill, J. Electroanal. Chem. Interfacial Electrochem. **122**, 321–326 (1981)
25. D. Yang, D. Bizzotto, J. Lipkowski, B. Pettinger, S. Mirwald, J. Phys. Chem. **98**, 7083–7089 (1994)
26. J.R. Lombardi, R.L. Birke, Acc. Chem. Res. **42**, 734–742 (2009)

Chapter 3
Simultaneous Measurement of Surface-Enhanced Raman Scattering and Conductance Using Mechanically Controllable Break Junction Technique

3.1 Introduction

In Chap. 2, I analysed the depolarisation behaviour of SERS in a metal nanogap structure. Depolarisation is strongly dependent on the adsorbed structure of the metal surface. Ensemble measurement of the adsorbed molecules and each metal structure showed fluctuation in the depolarisation SERS behaviour and the intensity ratio in the same spectrum. For further investigation of this depolarisation behaviour, it is important to fabricate a single molecular junction combined with spectroscopy of the single molecule. Up to now, vibrational spectroscopy on single molecules has been investigated in the field of the adsorbed molecules. In 1998, the first observation of an acetylene molecule was undertaken using scanning tunnelling microscopy (STM) [1]. Then, inelastic tunnelling microscopy [2–5], single molecular fluorescence [6] and STM light emission spectra, which have vibrational-induced structures [7], were measured to investigate single molecular vibrational spectroscopy.

A single metal atomic contact showed the specific conductance limited by the density of the electronic states characterised by free electron metals. In the case of Au, STM showed 1 G_0 (= 12.9 kΩ^{-1}) of single-atom conductance at the metallic nanowire [8–12]. The Au nanowire showed quantised conductance at the metallic junction. At the same time as the formation of the 1 G_0 conductance, single-atom contact was measured using an ultrahigh-vacuum electron microscope [9]. The quantised conductance in a confined small area of metal is shown as

$$G = n\left(\frac{2e^2}{h}\right) = nG_0 \quad (n = 1, 2, 3\ldots), \tag{3.1}$$

where e is the electron charge, h is Planck's constant, $G_0 = 2e^2/h = (12.9\,\mathrm{k\Omega})^{-1}$ is a unit of quantum conductance and n is the number of electronic states related to the

F. Nagasawa, *Studies on the Plasmon-Induced Photoexcitation Processes of Molecules on Metal Surfaces*, Springer Theses,
DOI 10.1007/978-4-431-56579-6_3

conductance. In a more general condition, the conductance of the metal atomic contact is expressed by the Landauer formula [13] as described in (3.2):

$$G = \left(\frac{2e^2}{h}\right) \sum_{i}^{n} T_i = G_0 \sum_{i}^{n} T_i \quad (n = 1, 2, 3 \ldots), \tag{3.2}$$

where T_i is the transmission probability of the i-th conductance channel. The conduction electron is affected by the scattering when there is transmittance in the metallic nanojunction. T_i denotes the transmission probability in the process of the scattering. If this junction is immersed in a solution of molecules, the adsorbed molecules combine to the gap of the junction during the breaking process. Then conductance of a metal/molecule/metal junction is measured. The system that causes a single molecular junction in the metal nanodimer structure shows the intrinsic conductance properties at the metal junction. If there are no intermolecular interactions, conductance originates from the addition of the conductance from consisted molecules. In other words, when the conductance is changed by multiple integrals, the smallest conductance unit indicates the formation of the single molecular junction.

In the junction, two facing triangle-like nanostructures show an intense EM field between the gap of junction via LSPR. The intense EM field makes it possible to measure SERS in a metal/molecule/metal junction. Until now, measurements have been performed at single-molecule junctions [14–17]. These systems had high sensitivity related to single-molecule detection owing to LSPRs. However, detailed features concerning the intensity ratio of the Raman spectrum have scarcely been analysed in relation to the molecular structure. Moreover, it is expected that polarised Raman measurement also illustrates the CT character of the molecule/metal interface.

In this chapter, simultaneous measurement of conductance and polarised SERS in relation to the molecular bridged metal nanojunction is performed. The conductance measurement proves the number of molecules, while polarised Raman measurement proves the orientation and CT character.

3.2 Experimental

The mechanically controllable break junction (MCBJ) sample was fabricated using electron-beam lithography (Fig. 3.1b). The bending substrate was a plate of Al of about 0.8 mm thick, 22 mm long and 11 mm wide. The 200-nm-thick Al_2O_3 film was formed on the Al plate by the anodic oxidation method. The oxide film (Al_2O_3) was used as an insulating film because the background of the Raman spectrum could be reduced compared to the system using organic insulating film. On this

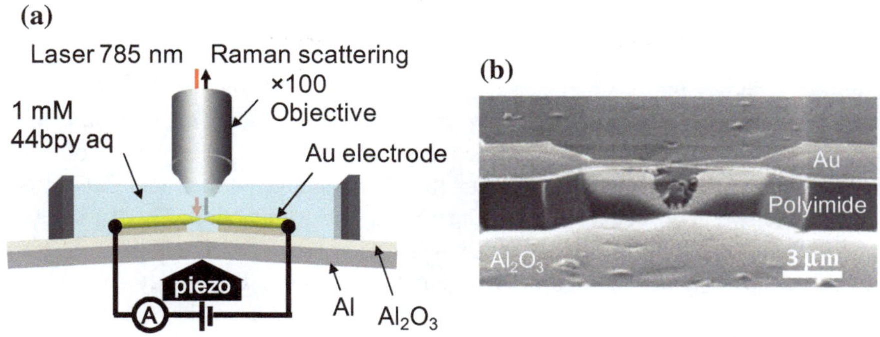

Fig. 3.1 Measurement system of the simultaneous measurement of **a** Raman and single-molecule conductance and **b** gap structure image by scanning electron microscope

substrate, a polyimide layer was spin coated with a thickness of 600 nm. The Au nanoelectrode was patterned on the substrate using electron beam lithography and the lift-off technique. Au (120 nm) and Cr (1 nm) were deposited using the electron beam evaporation technique. Subsequently, the polyimide underneath the Au nanoelectrode was removed by reactive ion etching using O_2/CF_4 plasma. A free-standing Au nanobridge was obtained. The whole assembly was mounted on a three-point bending mechanism consisting of a stacked piezo-element (NEC tokin) and two fixed counter supports. The water solution containing 1 mM 4,4'-bipyridine was dropped onto the Au nanobridge. By bending the substrate, the top surface of the substrate was expanded, and the Au nanobridge finally broke, forming a molecular junction in solution. By relaxing the bending of the substrate, the Au nanobridge could be re-established. The conductance measurement was carried out with a 20 mV bias voltage and a sampling rate of 1000 plot/s.

SERS measurement was performed using a homemade polarised Raman microprobe spectrometer. The excitation energy was 785 nm (1.58 eV) and the excitation power was 0.2 mW. Raman measurements were carried out in the backscattering configuration; scattering photons with parallel and perpendicular polarisation directions were collected simultaneously. The expanded NIR beam was focussed onto the sample using a water-immersion objective lens with 100× magnification and a numerical aperture of 1.0 (Fig. 3.1a). Raman image measurement was carried out by imaging Raman scattering at the excitation using an expanded NIR laser beam with a diameter of approximately 30 μm.

Raman spectra simulation were carried out using Gaussian 03, revision E.01, at the B3LYP level of DFT with the LanL2DZ basis set for Au atoms and 6–31 G** basis sets for other atoms. The $(Au)_2$ cluster was used in the calculations as a model of gold substrates. The vibration symmetry type was described by using the D_2 symmetry of the 4,4'-bipyridine molecule.

3.3 Results and Discussion

Figure 3.2 shows the conductance trace of single metal nanojunction at a single atomic junction region during breaking process of the contact. This trace illustrates the stepwise decrease with the time course. In this region, the plateau of 1 G_0 corresponds to a single atomic contact [8–12]. Moreover, the 2 and 3 G_0 trace indicate the two and three Au atoms at the gap of the Au junction. The Au junction immersed in 1 mM 4,4'-bipyridine aqueous solution showed not only the 1 G_0 trace but also a 0.01 G_0 trace in the process of junction breaking. Previously documented conductance has ranged from 1.6×10^{-4} to 0.01 G_0 [17–21]. This wide range may be due to differing experimental conditions (e.g. solvent, concentrations, bias voltage and sample treatment). I investigated the conductance of the single 4,4'-bipyridine molecule junction using the STM break junction technique under the same experimental conditions as those for the SERS measurements. Observation of conductance indicated the formation of a single molecular junction. In the present study, I observed significant modulations in both SERS intensity and the selectivity of the Raman vibrational bands that were coincident with current fluctuations in the single 4,4'-bipyridine molecular junction. Figure 3.2c shows the scanning electron microscope image of substrate. In the plateau of 0.01 G_0, a Raman image in the C–H vibrational mode (1018 cm^{-1}) was obtained, as shown in Fig. 3.2d. This reveals that the gap of the junction is main enhanced area.

Figure 3.3a shows the time course of the conductance around 0.01 G_0. Fluctuation of the conductance between 0.005 and 0.01 G_0 were observed. A conductance lower than 0.01 G_0 indicates the weak adsorption of the molecular bridging structure. The time course of conductance represents the switching between strong (1–6, 14–20 s) and weak adsorption strength (7–13 s). Raman spectra were acquired simultaneously (Fig. 3.3b). Strong, obvious Raman modes were measured throughout the time range. The additional Raman modes were measured in the higher conductance (= 0.01 G_0) region. Figure 3.3c (top) shows the typical Raman spectra in the region of lower conductance (<0.01 G_0). Raman

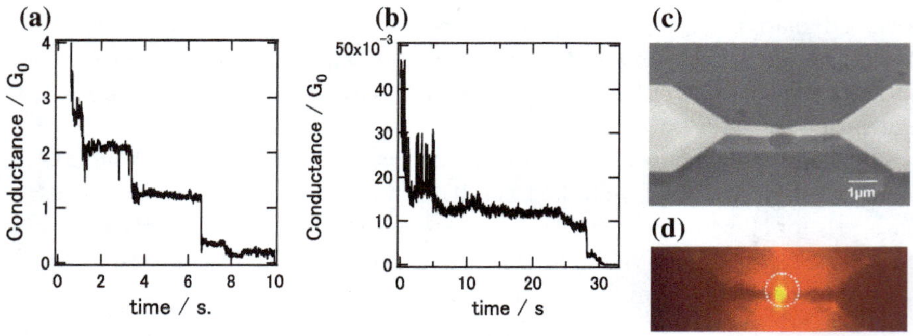

Fig. 3.2 Conductance trace of **a** the 1 G_0 area, **b** the 0.01 G_0 area, **c** SEM image, **d** Raman image at 1018 cm^{-1}

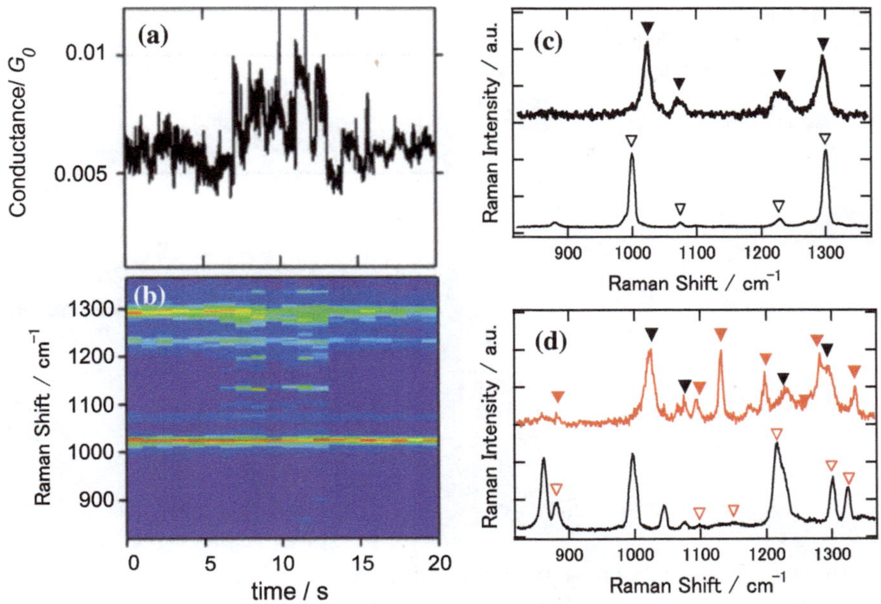

Fig. 3.3 Simultaneous measurement of **a** conductance and **b** Raman spectra, **c** (*top*) the Raman spectra of conductance <0.01 G_0 region and **c** (*bottom*) the polarised Raman spectrum of a single 4,4'-bipyridine crystal, **d** (*top*) conductance = 0.01 G_0 and **d** (*bottom*) polarised Raman spectrum

peaks at 1015, 1074, 1230 and 1297 cm^{-1} were measured (solid black arrows). Corresponding Raman modes with parallel polarisation molecular spectra are shown in Fig. 3.3c (bottom). These peaks were assigned to the totally symmetric a-mode. Figure 3.3d (top) shows the typical Raman spectra in the region of higher conductance (= 0.01 G_0). Additional Raman peaks at 880, 1095, 1132, 1198, 1282 and 1330 cm^{-1} were measured (solid red arrows). Additional peaks were assigned to the nontotally symmetric b_1 mode. The appearance of this additional nontotally symmetric mode was not explained by the EM field enhancement. As already shown in Chap. 2, it was predicted that the nontotally symmetric mode would be enhanced by the CT mechanism for SERS.

The CT mechanism was resonant Raman-like enhancement. In this case, I had to consider the energy state of the Au Fermi level, as well as the highest occupied molecular orbital (HOMO) and lowest unoccupied molecular orbital (LUMO) of the 4,4'-bipyridine molecule. When 4,4'-bipyridine molecules were bridged to the Au junction, hybridisation between the metal and molecule changed the LUMO (B_3 symmetry) energy to close to the Fermi level [22] (Fig. 3.4). In the present case, it is suggested that the CT from metal to LUMO was resonant to excitation energy. This energy level provided the x-direction-induced dipole moment, which was vertical to the pyridine rings. In the CT mechanism, term C in Lombardi et al.'s the model is represented by Eqs. (1.18) and (1.22) [23]:

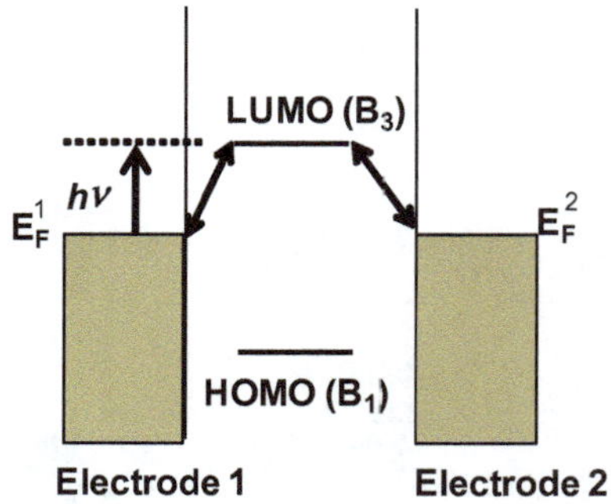

Fig. 3.4 Energy diagram of the resonant condition in the high-conductance region

$$\alpha_{\rho\sigma} = A_f + A_k + B + C, \tag{1.18}$$

$$C = -\frac{2}{\hbar}\sum_{K \neq I}\sum_{M > I} M^{\sigma}_{MK} M^{\rho}_{KI} h_{IM} \langle i|Q|f\rangle \frac{(\omega_{KI}\omega_{KM} + \omega^2)}{(\omega^2_{KI} - \omega^2)(\omega^2_{KM} - \omega^2)} \tag{1.22}$$

The resonant condition is predicted when $\omega \approx \omega_{KM}$. For the C-term to be nonvanishing, $\langle i|Q|f\rangle$ must be nonzero. In the fundamental mode, the usual selection rule becomes $f = i \pm 1$. In addition, it is required for C that neither h_{IK} nor components of M_{MK} and M_{KI} vanish. From the intensity borrowing the $I \rightarrow K$ transition, the vibrational mode can be arrowed. The Raman mode allowed by CT is described in Table 3.1 for each direction of CT, that is the adsorption orientation.

In the present system, the a-mode enhancement was measured in the lower conductance region. The totally symmetric a-mode was predicted by the EM field enhancement. If the CT resonance was also affected, the x-directed adsorbed orientation—that is, flat adsorption on the metal surface—was expected. The intensity ratio of the Raman spectrum predicted the z-directed adsorbed orientation in the junction. Therefore, the a-mode vibrational mode was induced by the EM effect. In contrast, a-mode and b_1-mode were coincidentally generated in the higher conductance region. The appearance of b_1-mode was expected to be produced in the y-directed adsorbed orientation by the CT mechanism (Table 3.1 and Fig. 3.5). The conductance showed the fluctuation of the junction from the breaking to the formation. The inclined orientation of the molecules situated in the gap provided the CT between the molecular and metal surface.

Table 3.1 Enhanced Raman Mode Selection Rule by the Charge Transfer Mechanism

Symmetry of LUMO	Induced dipole moment	Adsorbed orientation	Enhanced vibrational mode
B_3	$\mu_x(X)$	X	$a(\alpha_{xx}, \alpha_{yy}, \alpha_{zz})$
		Y	$b_1(\alpha_{xy})$
		Z	$b_2(\alpha_{xz})$

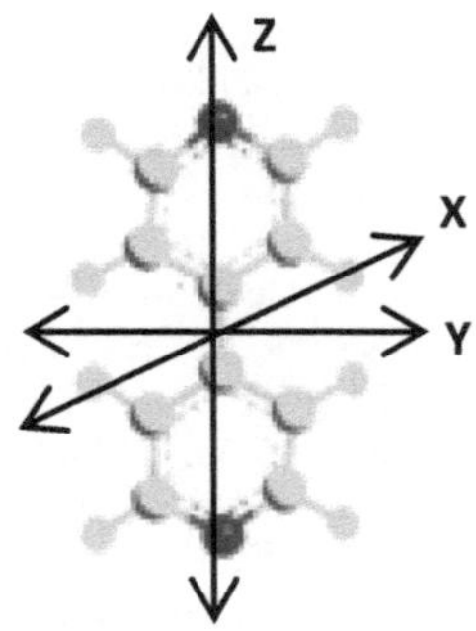

Fig. 3.5 The axis of 4,4'-bipyridine molecule

3.4 Conclusion

In this chapter, I successfully fabricated a single molecular environment to measure the Raman scattering, thereby measuring the conductance simultaneously. The conductance measurement clarified the number of molecules between the junction and strength of adsorption. Moreover, Raman measurement revealed presence of the CT from molecules to metal. The CT process provided the additional electronic resonant condition. The modulation of a small fraction of the adsorption structure at the interface strongly affected the process of the electric excitation.

References

1. B.C. Stipe, M.A. Rezaei, W. Ho, Science **280**, 1732–1735 (1998)
2. R.H.M. Smit, Y. Noat, C. Untiedt, N.D. Lang, M.C. van Hemert, J.M. van Ruitenbeek, Nature **419**, 906–909 (2002)
3. A. Troisi, M.A. Ratner, Nano Lett. **6**, 1784–1788 (2006)
4. J. Hihath, C.R. Arroyo, G. Rubio-Bollinger, N. Tao, N. Agraït, Nano Lett. **8**, 1673–1678 (2008)
5. M. Kiguchi, O. Tal, S. Wohlthat, F. Pauly, M. Krieger, D. Djukic, J.C. Cuevas, J.M. van Ruitenbeek, Phys. Rev. Lett. **101**, 046801 (2008)
6. X.H. Qiu, G.V. Nazin, W. Ho, Science **299**, 542–546 (2003)
7. Y. Uehara, S. Ushioda, Surf. Sci. **601**, 5643–5648 (2007)
8. J.I. Pascual, J. Méndez, J. Gómez-Herrero, A.M. Baró, N. García, V.T. Binh, Phys. Rev. Lett. **71**, 1852–1855 (1993)
9. H. Ohnishi, Y. Kondo, K. Takayanagi, Nature **395**, 780–783 (1998)
10. J.L. Costa-Krämer, N. García, H. Olin, Phys. Rev. B **55**, 12910–12913 (1997)
11. B. Ludoph, M.H. Devoret, D. Esteve, C. Urbina, J.M. van Ruitenbeek, Phys. Rev. Lett. **82**, 1530–1533 (1999)
12. K. Hansen, E. Lægsgaard, I. Stensgaard, F. Besenbacher, Phys. Rev. B **56**, 2208–2220 (1997)
13. R. Landauer, IBM J. Res. Dev. **1**, 223
14. D.R. Ward, N.J. Halas, J.W. Ciszek, J.M. Tour, Y. Wu, P. Nordlander, D. Natelson, Nano Lett. **8**, 919–924 (2008)
15. J.H. Tian, B. Liu, X. Li, Z.L. Yang, B. Ren, S.T. Wu, N. Tao, Z.Q. Tian, J. Am. Chem. Soc. **128**, 14748–14749 (2006)

16. X. Wang, Z. Liu, M.-D. Zhuang, H.-M. Zhang, X. Wang, Z.-X. Xie, D.-Y. Wu, B. Ren, Z.-Q. Tian, Appl. Phys. Lett. **91**, 101105/1–101105/3 (2007)
17. Z. Liu, S.-Y. Ding, Z.-B. Chen, X. Wang, J.-H. Tian, J.R. Anema, X.-S. Zhou, D.-Y. Wu, B.-W. Mao, X. Xu, B. Ren, Z.-Q. Tian, Nat. Commun. **2**, 1310/1–1310/6 (2011)
18. B. Xu, N.J. Tao, Science **301**, 1221–1223 (2003)
19. S.Y. Quek, M. Kamenetska, M.L. Steigerwald, H.J. Choi, S.G. Louie, M.S. Hybertsen, J.B. Neaton, L. Venkataraman, Nat. Nanotechnol. **4**, 230–234 (2009)
20. K. Horiguchi, S. Kurokawa, A. Sakai, J. Chem. Phys. **131**, 104703 (2009)
21. X.-S. Zhou, Z.-B. Chen, S.-H. Liu, S. Jin, L. Liu, H.-M. Zhang, Z.-X. Xie, Y.-B. Jiang, B.-W. Mao, J. Phys. Chem. C **112**, 3935–3940 (2008)
22. Z. Li, D.S. Kosov, J. Phys. Chem. B **110**, 9893–9898 (2006)
23. J.R. Lombardi, R.L. Birke, T. Lu, J. Xu, J. Chem. Phys. **84**, 4174–4180 (1986)

Chapter 4
Electronic Excitation of an Isolated Single-Walled Carbon Nanotube by Tuning Electrochemical Potential

4.1 Introduction

Localised surface plasmon resonance is an effective perturbation to transfer the energy of photons to electrons in the material/molecule interface. Electronic excitation is extremely important in controlling the orientation of the molecules situated on the metal surface and the electronic state at the interface. Single-walled carbon nanotubes (SWNTs) represent a promising tool as an analyte because of their electronic structures, which are well defined by the geometrical structure characterised by chirality [1, 2]. The density of the electronic states of SWNTs, which can be controlled by chemical/electrochemical doping, is also a unique measure for the determination of potential diagrams of the system.

SWNTs have intrinsic excitation properties [3–13]. An SWNT can be viewed as a strip cut from an infinite graphene sheet that is rolled up seamlessly to form a tube. The chirality determines the diameter and the chiral angle. The diameter and helicity of a SWNT are defined by the roll-up vector $C_h = n\mathbf{a}_1 + m\mathbf{a}_2 \equiv (n, m)$, which connects crystallographically equivalent sites on this sheet. Here, $\mathbf{a}_1$ and $\mathbf{a}_2$ are the graphene lattice vectors, and n and m are integers. These structures of SWNT show individual van Hove singularities in the electronic density of state (EDOS) because of the one-dimensional (1D) structure of SWNTs. Electronic absorption in symmetrical vHS ($\pm$ E_i) against the Fermi level is allowed in parallel absorption to the nanotube axis. In the perpendicular polarisation, $E_i \rightarrow E_{i\pm 1}$ is allowed. When this E_{ii} transition is resonant with the excitation energy, the resonant Raman intensity becomes active. RRS is widely used for the sample evaluation and characterisation of carbon nanotubes because it is a noncontact, nondestructive measurement [13–15]. Moreover, these techniques can operate at room temperature and in air.

The low frequency features (100–400 cm^{-1}) show the radial breathing mode (RBM). This mode represents the vibration of carbon atoms in a radial direction in relation to the nanotube axis. The graphite-like band (G-band) in carbon nanotubes

F. Nagasawa, *Studies on the Plasmon-Induced Photoexcitation Processes of Molecules on Metal Surfaces*, Springer Theses,
DOI 10.1007/978-4-431-56579-6_4

is directly related to the G-band in graphite, which is identified with an in-plane tangential optical phonon involving the stretching of the bond between the two atoms in the graphene unit cell. In the region of the 1500 cm^{-1} G-band is measured in the Raman spectrum. Because of the highly asymmetric feature of SWNTs, the G-band is split into six bands, referred to as A^+, A^-, E_1^+, E_1^-, E_2^+ and E_2^- modes. These Raman intensity ratios of the spectrum reflect the orientation to the light polarisation [16]. In the metal nanogap that causes the anisotropic EM field, the intensity ratios of the G-band are assigned to the direction in the gap. From the Raman spectrum, I could determine both the singularity of SWNTs and the orientation of the SWNTs against the LSPR EM field direction. In the region of 1300 cm^{-1}, the D-band was measured, which originated from the disorder-induced mode.

For the purpose of estimation of the SWNTs, an electrochemical environment was adopted to modulate the doping density. Bard et al. presented the first electrochemical study of a SWNT deposited on Pt or Au electrodes [17]. The first observation concerning an individual SWNT was presented in 2005 by Heller et al. [18]. In 1997, pioneering redox-doped SWNT Raman investigation was performed (without using electrochemistry) [19]. Raman spectra of bundle SWNTs were acquired in the electrochemical environment in relation to chemical/electrochemical doping. In 2000, Kavan et al. studied electrochemistry in aqueous electrolyte solution [20] and Fischer et al. studied it in aprotic electrolyte solution [21]. The absolute potential of the Fermi level in an isolated nanotube was determined by Okazaki et al. [22–24].

It is known that the resonant Raman profile reflects the vHSs. SWNTs have quantum capacitance properties that are observed in the electrochemical environment [17, 18, 25–27]. This characteristic maintains the EDOS when the electron filling of the conduction band and depletion of the valence band occur. Only the Fermi level is tuned in the electrochemical environment [22, 28, 29]. Spectroelectrochemistry, particularly in terms of Raman measurement, has provided fruitful information about SWNTs. The frequency of RBM is maintained during the tuning of the electronic density of the state. The intensity of RBM and the G-band are modulated by the depletion/filling of valence/conduction band. The frequency of the G-band and D-band vary according to the electrochemical potential. The measurement of Raman scattering combined with electrochemistry is supposed to clarify the EDOS in the isolated SWNT level [30–32].

To clarify the electric excitation behaviour in the LSPR-induced EM field, polarised Raman measurement combined with electrochemistry is extremely important. However, the combination of the polarised Raman measurement in the electrochemical environment has not yet been performed. In this chapter, I measure the polarised Raman measurement performed in the electrochemical environment using a metal nanodimer array. By confirming the singularities concerning SWNTs, I analyse the G-band intensity ratios by changing the electrochemical potential. Then, I analyse the electronic excitation of the SWNTs.

4.2 Experimental

In the experimental approach, Au dimer arrays were fabricated on indium tin oxide (ITO) glass as a working electrode using the AR-NSL technique. The sample of carbon nanotubes was prepared using the alcohol catalytic chemical vapour deposition method. The SWNT sample was monodispersed in N-methyl-2-pyrrolidone solution. Individual SWNTs were added to the Au-NSL surface by dropwise introduction. Polarised Raman measurement (λ_{ex} = 785 nm) was acquired in the backscattering configuration using an immersion lens (×100) under electrochemical potential control. The aprotic electrolyte of ionic liquid of N, N,N-trimethyl-N-propylammonium bis (trifluoromethanesulfonyl) amide (TMPA-TFSA) was used in the electrochemical measurement. The solution was dehydrated overnight at 110 °C. The sample was then situated in a vacuum oven overnight for deoxidation. The reference electrode was Ag/0.05 M Ag^+ TMPA-TFSA (4.92 V vs. vacuum level) and the counter electrode was a Pt plate.

4.3 Results and Discussion

Figure 4.1a shows the typical resonant Raman spectrum (excitation wavelength λ_{ex} = 785 nm). In the region of 100–300 cm^{-1}, these modes are assigned to the RBM. Figure 4.1b shows the typical surface-enhanced resonant Raman spectrum at the metal nanogap. The single RBM mode obtained at 241 cm^{-1} is assigned (8, 6) semiconducting nanotubes. Figure 4.1c shows the Raman intensity profile as a function of the incident polarisation angle to the dimer long axis. The dotted line presents $\cos^2\alpha$ correlation, which is concerned with the LSPR EM field correlation. This represents the SWNT affecting the EM field of LSPR in the gap. As a result, it was found that the SERRS measurement of isolated SWNT was achieved.

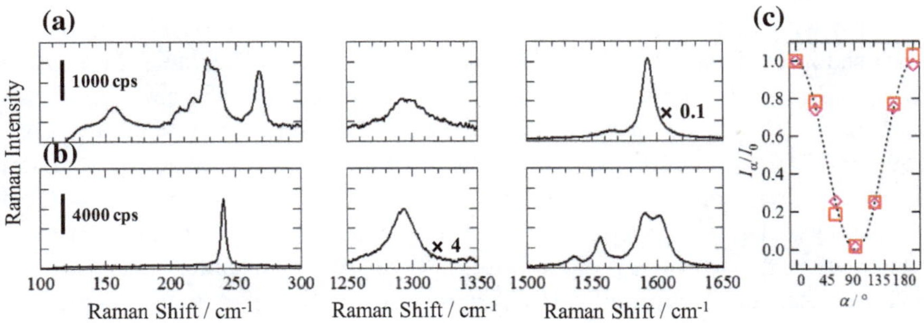

Fig. 4.1 **a** Resonant Raman scattering (RRS) spectrum, **b** surface-enhanced resonant Raman scattering (SERRS) spectrum and **c** SERRS intensity profile as a function of the incident polarisation angle to the dimer long axis. Excitation power I_{ex} = 2 mW (**a**) and I_{ex} = 0.2 mW (**b**). Exposure time t_{ex} = 100 s (**a**), exposure time t_{ex} = 1 s (**b**)

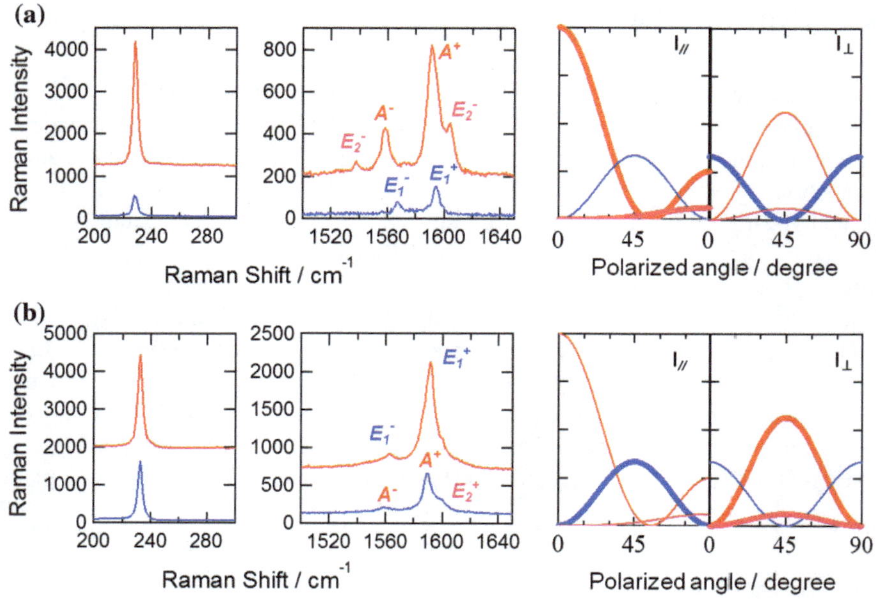

Fig. 4.2 Polarised Raman spectra of different sites, excitation power I_{ex} = 0.2 mW, exposure time t_{ex} = 1 s: **a** ω_{RBM} = 228 cm^{-1}, d = 1.04 nm, semiconducting (10, 5) nanotube and **b** ω_{RBM} = 233 cm^{-1}, d = 1.00 nm, semiconducting (11, 3) nanotube

Surface-enhanced Raman scattering was acquired at each polarised configuration (Fig. 4.2a, b). The RBM modes were obtained at ω_{RBM} = 228 cm^{-1} and ω_{RBM} = 233 cm^{-1}. Both were assigned to semiconducting nanotubes. As shown in Fig. 4.2a, G-bands were acquired at 1538 cm^{-1} (E_2^-), 1558 cm^{-1} (A^-), 1591 cm^{-1} (A^+) and 1603 cm^{-1} (E_2^+) in the parallel polarisation configuration (red solid line) and 1567 cm^{-1} (E_1^-) and 1594 cm^{-1} (E_1^+) in the perpendicular polarisation configuration (blue solid line). In the polarised Raman activity correlation at each incident polarisation angle (α) in Fig. 4.2a (right side), A and E_2 modes were obtained without E_1 mode at α = 90° against the dimer long axis. In the parallel polarisation configuration at α = 90°, only E_1 mode was observable. In a similar fashion, Fig. 4.2b shows that G-bands were acquired at 1564 cm^{-1} (E_1^-) and 1593 cm^{-1} (E_1^+) under parallel polarisation configuration and 1560 cm^{-1} (A^-), 1591 cm^{-1} (A^+) and 1600 cm^{-1} (E_2^+) under perpendicular polarisation configuration. This mode appearance at each polarisation configuration showed α = 45° at the nanogap. The A and E_2 modes were observed in the parallel transition to the nanotube axis, and the E_1 mode was observed in the parallel transition. The polarised Raman spectrum of SWNTs was proven to involve an anisotropic transition around the nanotube axis. With the purpose of comparison with RRS and SERRS, I obtained in the same condition of chirality and polarisation angle α. Figure 4.2a, b shows ω_{RBM} = 233 cm^{-1} and the (11, 3) semiconducting nanotube (d = 1.00 nm). The intensity was affected by the signal enhancement in each

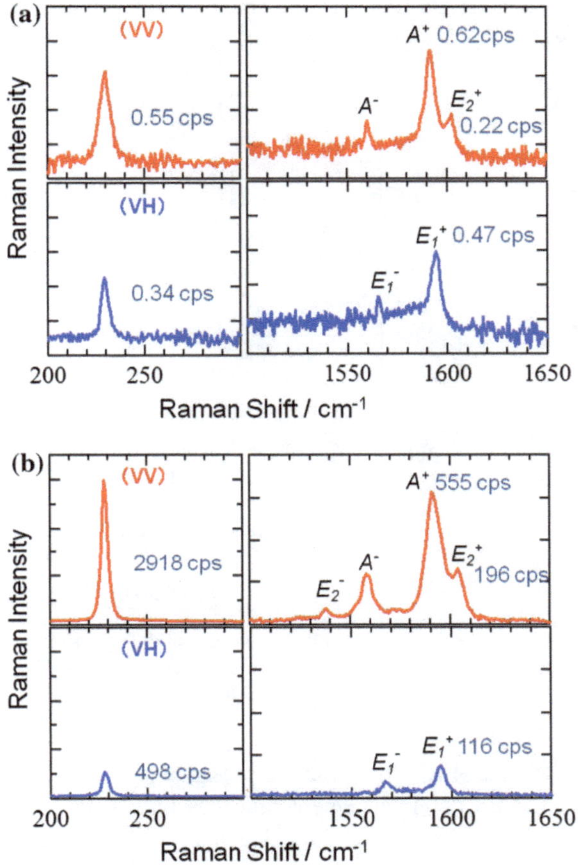

Fig. 4.3 **a** Polarised resonant Raman spectrum and **b** polarised surface-enhanced resonant Raman spectrum. $\omega_{RBM} = 228\ cm^{-1}$, d = 1.04 nm, semiconducting (10, 5) nanotube (**a**, **b**)

configuration. The RRS showed the G-bands at 1558 cm^{-1} (A^-), 1591 cm^{-1} (A^+) and 1603 cm^{-1} (E_2^+) in the parallel polarisation configuration (red solid line) and 1567 cm^{-1} (E_1^-) and 1594 cm^{-1} (E_1^+) in the perpendicular polarisation configuration (blue solid line), like SERRS (Fig. 4.3b). The intensity ratio of these G-band features indicated that $\alpha = 45°$ against the tube axis. The SERRS spectrum showed a similar depolarisation feature to the resonant Raman spectra. In the resonant condition with the LSPRs, SERRS spectra showed intense Raman scattering because of a twofold enhancement mechanism. The enhancement appeared in the parallel polarisation configuration in the LSPR direction. Although the perpendicular polarisation configuration was hardly responsible for the enhancement, polarised SERRS spectrum showed an enhanced signal to a certain extent. This feature was examined according to the intensity ratio between RRS and SERS. From the intensity ratio in the parallel polarisation configuration, in addition to the confinement of the enhanced field $\sim 10^{3-4}$, the enhancement factor at the incident and scattering polarisation was estimated to be about $0.8–2.7 \times 10^3$. By using this factor in the perpendicular polarisation configuration, the intensity ratio between

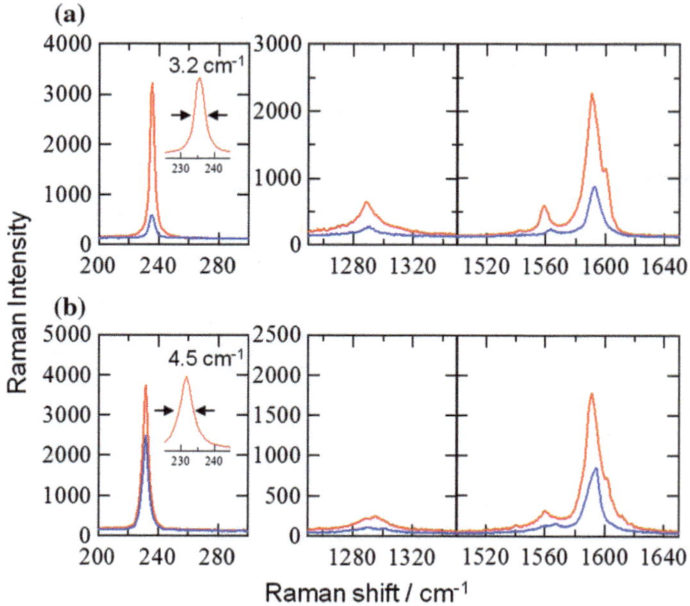

Fig. 4.4 Polarised resonant Raman spectra at different site: parallel polarisation configuration (*red line*) and perpendicular polarisation configuration (*blue line*). ω_{RBM} = 228 cm^{-1}, d = 1.04 nm, semiconducting (11, 3) nanotube (**a**, **b**)

RRS and SERRS showed the presence of additional enhancement (1.2–3.9 $\times$ 10^2) to the LSPR enhancement that was only generated in the incident process. It is suggested that the additional enhancement at the perpendicular polarisation was caused by the electronic excitation near the metal surface.

Electrochemical potentials were tuned in the same substrate in ionic liquid for the purpose of measuring the electronic state in SWNTs. Figure 4.4 shows the polarised SERRS spectra with different half bandwidths. Figure 4.4a shows the bandwidth at 3.2 cm^{-1} and (b) illustrates that at 4.5 cm^{-1}. These differences come from the σ–π coupling of nanotube or doping density in the confined area on the SWNTs [23]. Figure 4.5 shows the polarised SERRS spectra under the electrochemical potential control. The RBM and G-band intensity increased and decreased under the potential control. This tendency is suggested to be related to the depletion/filling of valence/conduction band.

Figure 4.5a shows the intensity modulation without the band shift. In addition, Fig. 4.5b shows the G-band intensity modulation with band shift in the parallel polarisation configuration, and Fig. 4.5c shows the similar modulation in the G-band intensity and band shift in the perpendicular polarisation configuration. The intensity and band-shift profile as a function of the potential are shown in Fig. 4.6. The calculated EDOS values are shown in Fig. 4.6a. The maximum Raman intensity was presented according to the Fermi level. Moreover, the Raman intensity presented different tendencies at the $\pm$ E_1 and $\pm$ E_2 levels. Frequency shifts were increased

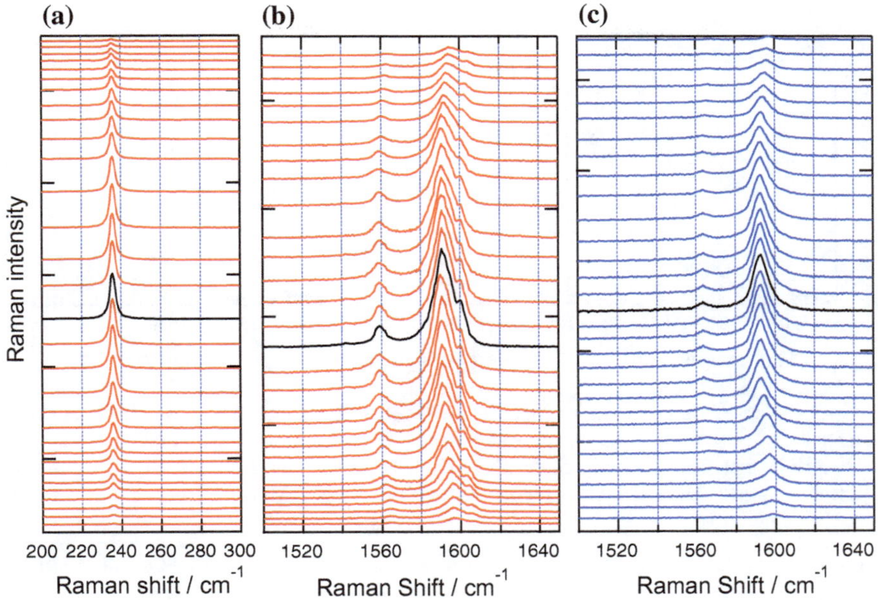

Fig. 4.5 Polarised SERRS spectra under the electrochemical potential control of (**a**, **b**) parallel polarisation configuration and perpendicular polarisation configuration (**c**). RBM linewidth $\Gamma_{RBM} = 4.5\ cm^{-1}$. The potentials are +1.5 V to −1.5 V versus Ag/Ag^{+} from bottom to top in 0.1 V increments

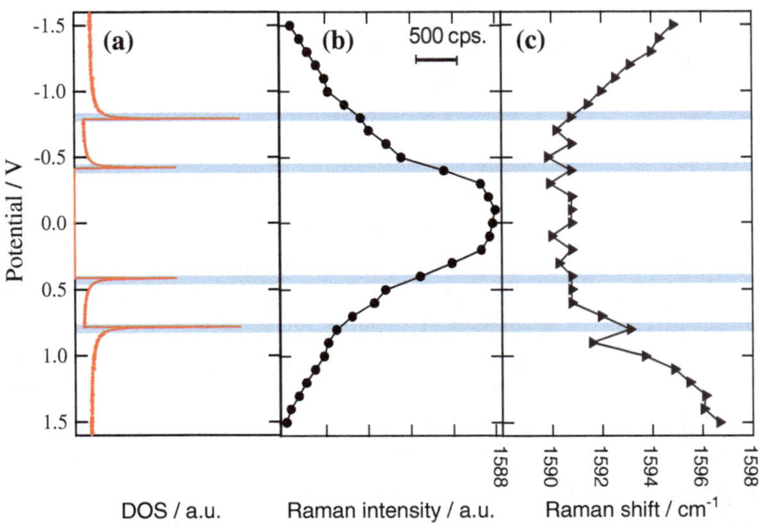

Fig. 4.6 **b** Raman intensity profile **c** Raman frequency profile and **a** the calculated electronic density of state (EDOS) of the corresponding nanotube

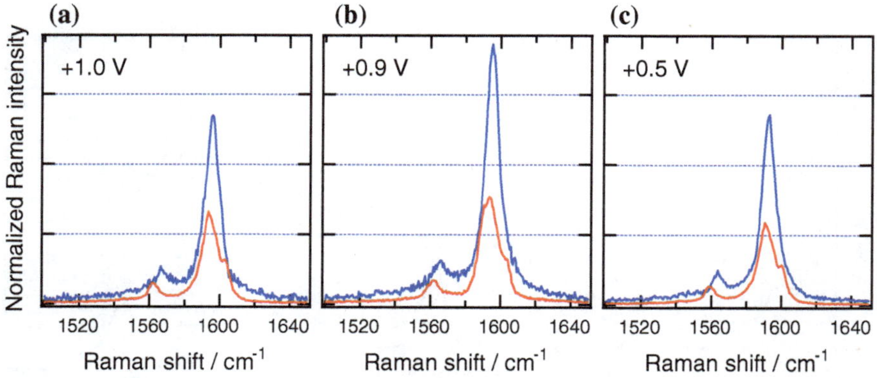

Fig. 4.7 Polarised Raman spectra normalised by the RBM band intensity at each electrochemical potential: **a** +1.0 V, **b** +0.9 V, **c** +0.5 V. Parallel polarisation configuration (*red line*), perpendicular polarisation configuration (*blue line*). Exposure time $t_{ex} = 1$ s and $I_{ex} = 0.2$ mW. Electrolytes were TMPA-TFSA; each potential versus Ag/Ag^{+} TMPA-TFSA

around the $\pm$ E_2 levels. In the more positive potential of $+E_1$ and $+E_2$, the resonant Raman intensity was attenuated because of the electron depletion of the valence band. In the more negative potentials of $-E_1$ and $-E_2$, the resonant Raman intensity was also attenuated because of the electron filling of the conduction band.

In the narrow RBM band width in Fig. 4.4a, fluctuations of the depolarisation ratio were observed near the vHS (Fig. 4.7). The electron filling of the valence band enhanced the E_1 mode in the perpendicular polarisation configuration. The E_1 mode is known to be the $E_{n,n\pm1}$ transition in the RRS spectrum. The E_1 band intensity ratio increased at +0.9 V, which is near to the $+E_2$ band. Additional transitions of $E_{2\rightarrow1}$ are suggested to enhance the perpendicular polarisation signal.

4.4 Conclusion

In this chapter, I obtained the isolated SWNT Raman spectra using the Au nanodimer arrays. The Au dimer structured on a conductive substrate allowed in situ electrochemical Raman measurement to be carried out. The SERRS observed in the present system showed depolarised behaviour, especially in the specific G-band mode. The G-band intensity, especially in the E_1 mode, was enhanced in the perpendicular polarisation configuration near vHS. This tendency demonstrated that the additional transition on the vertical transition ($E_{2\rightarrow1}$) affected the perpendicular Raman signal enhancement. The electrochemical potential dependence suggests that a highly localised anisotropic EM field excites the specific transition of $E_{2\rightarrow1}$ or $E_{1\rightarrow2}$, which results in the formation of excited electrons and holes at more negative and positive potentials than those formed at normal light illumination in SWNTs. The polarised surface-enhanced resonant Raman spectra clarified these characteristic transitions of the electronic states under electrochemical potential control.

References

1. M.S. Dresselhaus, G. Dresselhaus, R. Saito, Phys. Rev. B **45**, 6234–6242 (1992)
2. G.G. Samsonidze, A. Grüneis, R. Saito, A. Jorio, A.G. Souza Filho, G. Dresselhaus, M. Dresselhaus, S. Phys. Rev. B **69**, 205402 (2004)
3. R. Saito, M. Fujita, G. Dresselhaus, M.S. Dresselhaus, Appl. Phys. Lett. **60**, 2204–2206 (1992)
4. G.G. Samsonidze, A.R. Saito, D.A. Jorio, E.M.A. Pimenta, E.A.G. Souza Filho, F.A. Grüneis, D.G. Dresselhaus, M. Dresselhausa, S. J. Nanosci. Nanotechnol. **3**, 431–458 (2003)
5. S.J. Tans, M.H. Devoret, H. Dal, A. Thess, R.E. Smalley, L.J. Geerligs, C. Dekker, Nature **386**, 474–477 (1997)
6. G.G. Samsonidze, R. Saito, N. Kobayashi, A. Grüneis, J. Jiang, A. Jorio, S.G. Chou, G. Dresselhaus, M.S. Dresselhaus, Appl. Phys. Lett. **85**, 5703–5705 (2004)
7. R. Saito, G. Dresselhaus, M.S. Dresselhaus, Phys. Rev. B **61**, 2981–2990 (2000)
8. N.P. Valentin, New J. Phys. **6**, 17 (2004)
9. H. Telg, J. Maultzsch, S. Reich, F. Hennrich, C. Thomsen, Phys. Rev. Lett. **93**, 177401 (2004)
10. C. Fantini, A. Jorio, M. Souza, M.S. Strano, M.S. Dresselhaus, M.A. Pimenta, Phys. Rev. Lett. **93**, 147406 (2004)
11. V.N. Popov, L. Henrard, P. Lambin, Phys. Rev. B **72**, 035436 (2005)
12. H. Kataura, Y. Kumazawa, Y. Maniwa, I. Umezu, S. Suzuki, Y. Ohtsuka, Y. Achiba, Synth. Met. **103**, 2555–2558 (1999)
13. M.S. Dresselhaus, G. Dresselhaus, R. Saito, A. Jorio, Phys. Rep. **409**, 47–99 (2005)
14. A. Jorio, A.G. Souza Filho, G. Dresselhaus, M.S. Dresselhaus, A.K. Swan, M.S. Ünlü, B.B. Goldberg, M.A. Pimenta, J.H. Hafner, C.M. Lieber, R. Saito, Phys. Rev. B **65**, 155412 (2002)
15. C. Fantini, M.A. Pimenta, M.S. Strano, J. Phys. Chem. C **112**, 13150–13155 (2008)
16. A. Jorio, Phys. Rev. Lett. **90**, 107403 (2003)
17. Liu, C. y.; Bard, A. J.; Wudl, F.; Weitz, I.; Heath, J. R. *Electrochem. and Solid-State Lett.* **1999,** 2, 577–578
18. I. Heller, J. Kong, H.A. Heering, K.A. Williams, S.G. Lemay, C. Dekker, Nano Lett. **5**, 137–142 (2005)
19. A.M. Rao, P.C. Eklund, S. Bandow, A. Thess, R.E. Smalley, Nature **388**, 257–259 (1997)
20. L. Kavan, P. Rapta, L. Dunsch, Chem. Phys. Lett. **328**, 363–368 (2000)
21. A.S. Claye, N.M. Nemes, A. Jánossy, J.E. Fischer, Phys. Rev. B **62**, R4845–R4848 (2000)
22. K.-I. Okazaki, Y. Nakato, K. Murakoshi, Phys. Rev. B **68**, 035434 (2003)
23. K. Murakoshi, K.-I. Okazaki, Electrochim. Acta **50**, 3069–3075 (2005)
24. K.-I. Okazaki, Y. Nakato, K. Murakoshi, Surf. Sci. **566–568**, 436–442 (2004)
25. I. Heller, J. Kong, K.A. Williams, C. Dekker, S.G. Lemay, J. Am. Chem. Soc. **128**, 7353–7359 (2006)
26. E. Frackowiak, F. Béguin, Carbon **39**, 937–950 (2001)
27. E. Frackowiak, F. Béguin, Carbon **40**, 1775–1787 (2002)
28. M.J. O'Connell, E.E. Eibergen, S.K. Doorn, Nature Mat. **4**, 412–418 (2005)
29. Y. Tanaka, Y. Hirana, Y. Niidome, K. Kato, S. Saito, N. Nakashima, Angew. Chem. Int. Ed. **48**, 7655–7659 (2009)
30. M. Kalbac, H. Farhat, L. Kavan, J. Kong, M.S. Dresselhaus, Nano Lett. **8**, 3532–3537 (2008)
31. M. Kalbac, Y.-P. Hsieh, H. Farhat, L. Kavan, M. Hofmann, J. Kong, M.S. Dresselhaus, Nano Lett. **10**, 4619–4626 (2010)
32. M. Kalbac, H. Farhat, L. Kavan, J. Kong, K.-I. Sasaki, R. Saito, M.S. Dresselhaus, ACS Nano **3**, 2320–2328 (2009)

Chapter 5
Raman Enhancement via Polariton States Produced by Strong Coupling Between Localised Surface Plasmons and Dye Excitons in Metal Nanodimers

5.1 Introduction

The character of the plasmon-induced photoexcitation process is modulated by the interaction between metal and molecules. Moreover, the energy of photoexcitation process is limited by the intrinsic energy of the materials. The search for active interactions between metals and molecules is necessary to determine the general versatility and wide-range use of photoexcitation. Intensive investigations on the interactions between the electronic states in organic molecules and plasmons have been undertaken. In particular, it was found that when an emitter situated near the metal surface interacts with the plasmon resonance, the spontaneous emission rate is modulated and the energy transfer rate is changed [1, 2]. Most studies, however, have been carried out in relation to the weak-coupling regime. In contrast, pioneering studies have proved that organic dye molecules with high oscillator strengths form strong-coupling regimes under excitation by plasmon [3–11]. The coupling strength can be controlled via the plasmon energy and its linewidths. Furthermore, it has been found that hybrid states produced by strong coupling modulate the optical absorption and emission properties [3–5], the lifetimes of the excited states [6] and the chemical reactivity [10] of the systems. Although many reports have observed strong coupling produced by SPs [3, 6–9], application of the LSPR remains a challenging subject [12–16].

The photoexcitation characteristics of materials have been investigated using SERRS. Electronic excitation of molecules due to localised EM field leads to additional enhancement resulting from the optical absorption resonance in the SERRS process. Previously documented studies have found that the absorption spectra of chromophores that were adsorbed to Au or Ag plasmonic nanoparticles can be strongly perturbed [17–22]. This perturbation of the electronic transition affects the SERRS profiles in ways that differ from those predicted based on simple enhancement of the local EM field [23–25]. Brus et al. described the SERRS effect as involving transition dipole coupling between the chromophore and the metal

F. Nagasawa, *Studies on the Plasmon-Induced Photoexcitation Processes of Molecules on Metal Surfaces*, Springer Theses,
DOI 10.1007/978-4-431-56579-6_5

[26, 27]. In addition, Mayer showed a theoretical simulation of the SERRS conditions that included resonance coupling between the molecular electronic transition and the optical transition of the nanoparticles [28]. Despite these attempts to study the Raman characteristics when the system achieves hybridised states [29], the Raman features in the strong-coupling regime when using LSPR have not yet been clarified.

In this work, I report on the effects of Raman scattering on the formation of nanodimer–organic dye molecule complexes that achieve strong coupling between the plasmons of the nanodimer structure and the excitons of the adsorbed dye molecules.

5.2 Experimental

Ag dimer arrays were prepared by AR-NSL using repeated vapour depositions with different angles onto a PS particle (diameter d = 350 nm) monolayer prepared on a glass substrate. After the first deposition at 0° in the angle between the surface normal and the Ag deposition beam, the second deposition was carried out at the different angles, θ, varying from 19° to 23°. The Ag-NSL substrate was obtained after the removal of PS particles (Fig. 5.1a–d). The dye used was 1,1',3,3,3',3'-hexamethylindotricarbocyanine perchlorate (HITC, CAS No. 16595-48-51, Fig. 5.1e).

An HITC-covered Ag-NSL substrate (Ag@HITC) was fabricated by dropping 100 µl of 10 µM HITC aqueous solution. After drying the HITC solution, the extinction spectrum was acquired. For Raman measurement, the Ag-NSL substrate was rinsed to remove residual HITC. The extinction spectra of the metal dimer array in the visible-NIR region were recorded using a multichannel spectrometer (MCPD-2000, Ohtsuka Electronics). The structure of the dimer on the glass substrate was inspected using an AFM (Nanoscope-IIIa, Digital Instruments) in air. A home-made polarised Raman microscope spectrometer was specially modified for NIR laser light (r_{ex} = 785 nm, 1.58 eV)[chem. comm] with 100× objective

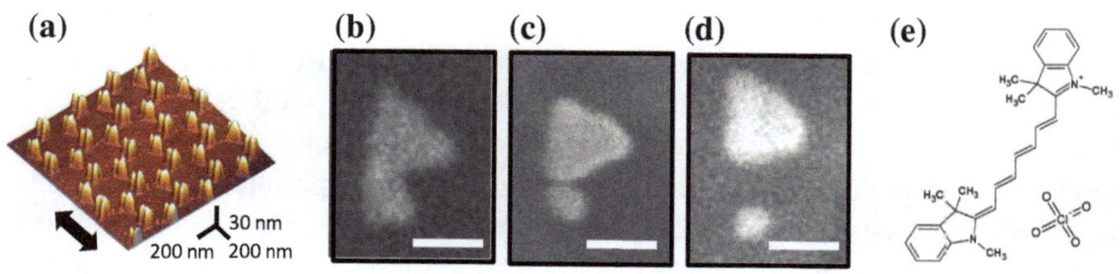

Fig. 5.1 **a** Atomic force microscope (AFM) image of the Ag-NSL substrate (d = 350 nm); **b**–**d** SEM image of the Ag-NSL substrate, **e** second deposition angle θ = 19° (**b**), 21° (**c**) and 23° (**d**), scale bar indicates 50 nm; HITC molecular structure (**e**)

lens; Olympus, 1.0 NA). Raman measurements were carried out at the backscattering configuration collecting the scattering photons with the parallel and perpendicular polarisation directions simultaneously.

5.3 Results and Discussion

The coupling strength was tuned using nanodimers with gap-mode plasmon wavelengths varying from 600 nm (2.07 eV) to 850 nm (1.46 eV) through the exciton energy of the organic dye molecule at 750 nm (1.65 eV). The plasmon energy was changed by varying the distance between the Ag dimers, which were prepared by AR-NSL on a glass substrate (surface area = 1 cm^2) [30–33]. The estimated gap distance of a tip-to-side triangular dimer structure with an approximate side length of 50 nm and height of 30 is a few nm when the dimer structure shows an extinction maximum at 785 nm (1.58 eV) for Raman excitation laser light polarised parallel to the long axis of the dimer structure (Fig. 5.1a–d). This gap distance was varied from 0 to <18 nm for changing gap-mode plasmon wavelengths from 850 to 600 nm [30, 33, 34]. Polarised Raman measurements provide information about the contributions of the hybridised states at the metal dimer gap to the Raman enhancement. Figure 5.1a shows a typical AFM image of the Ag-NSL substrate. The directions of the long axis of the individual dimers were well aligned on the glass substrate. The dye used in the present system is HITC (Fig. 5.1e). Figure 5.2a shows the absorption spectrum of 10 mM HITC aqueous solution. The extinction spectrum was split into two bands at the incident light at parallel polarisation along the dimer long axis, as shown in Fig. 5.2b. At the perpendicular polarisation along the dimer long axis, the extinction spectrum showed a peak at the dye exciton energy. For the evaluation of the split peak of extinction spectrum at parallel incident light, I checked the dependence on the amount of the dye layer. Figure 5.3a shows a bare Ag-NSL substrate (top column), demonstrating a bare plasmon peak at 694 nm. The HITC deposited on Ag-NSL substrate (Ag@HITC) was prepared by dropping 10 μM HITC aqueous solution onto the Ag-NSL substrate. After drying of the HITC solution, extinction spectra were acquired, and the resulting spectra are shown in the bottom column of Fig. 5.3a. The spectral change was observed as a split in the peak of the extinction. The peak wavelength can be estimated via deconvolution of the observed spectrum into two distinct peaks. The separation of the peaks increased with the amount of HITC used, as shown in Fig. 5.3a, from top to bottom.

Figure 5.3b shows the energy estimated from the peak separation as a function of a square root of the integrated extinction of the glass substrate with the same amount of the dye molecule as Ag@HITC. The splitting energy after the first deposition of 100 μl of a 10 μM HITC aqueous solution was observed 0.26 eV. After repeated deposition with a total of 600 μl of a 10 μM HITC aqueous solution, the splitting energy reached a maximum value of 0.39 eV. The dependence of the values of the splitting energy on the square root of the integrated extinction of the

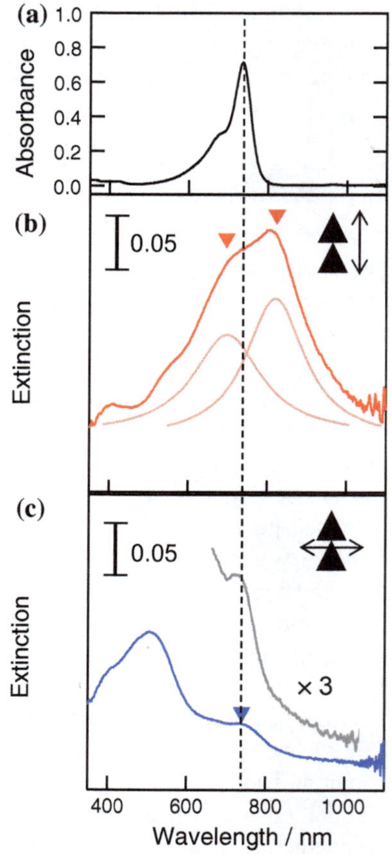

Fig. 5.2 Absorption spectra of **a** 10 mM HITC aqueous solution, **b**, **c** the extinction spectrum, incident polarisation parallel to the longer dimer axis (**b**) and shorter dimer axis(**c**)

dye molecules showed linear correlation, as illustrated in Fig. 5.3b. The observed energy splitting could be attributed to the strong coupling between excitons in the dye molecules at the gap and localised plasmons of the metal nanodimer. When plasmons and excitons achieve strong coupling, the extinction spectrum shows two extinction peaks, with the energy separation characterised as

$$\Delta E_{\mathrm{Rabi}} = \sqrt{(\hbar\Omega_{\mathrm{R}})^2 - (\gamma_{\mathrm{LSP}} - \gamma_{\mathrm{HITC}})^2} \quad (5.1)$$

where ΔE_{Rabi} is the energy separation between extinction peaks; $\hbar\Omega_R$ is the coupling energy due to Rabi splitting; and γ_{LSP} and γ_{HITC} are LSP resonance and molecular resonance linewidths, respectively, [16]. The observed splitting energy of 0.39 eV may reflect the coupling of the LSPR and the excitons of the HITC molecules. In the theoretical calculation, the Rabi splitting occurs when $\hbar\Omega_R < (\gamma_{\mathrm{LSP}} - \gamma_{\mathrm{HITC}})/2^{3^4}$. Based on the values of the relative resonance linewidths (γ_{LSPR} = 0.14 eV and γ_{HITC} = 0.05 eV), the observed splitting energy agrees with

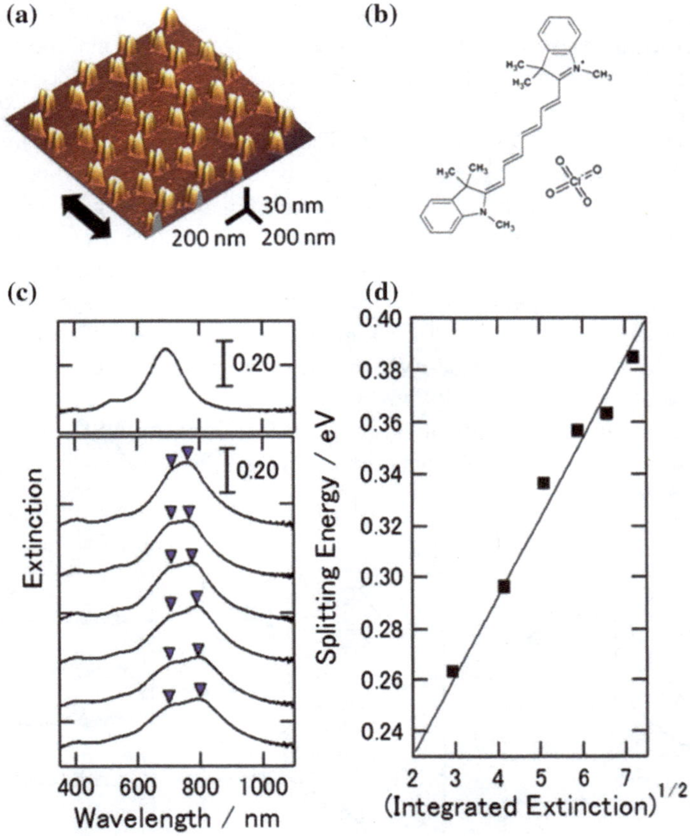

Fig. 5.3 **a**, *upper column* Polarised extinction spectrum of the bare Ag-NSL substrate, with the parallel polarisation to the long axis of the dimers; **a**, *bottom column* the extinction spectrum of HITC deposited on Ag-NSL with increasing HITC coverage from top to bottom; **b** the corresponding energy of Rabi splitting as a function of the square root of the integrated extinction of the dye molecules

the values of the strong-coupling regime in previously documented results [16] and theoretical studies [34]. Although the value of the mode volume of the LSPR field produced by the metal nanodimer structure used here is difficult to define, the localisation of the photon field of the present system at less than a few nm^3 in space suggests the contribution of a relatively small number of molecules at the gap to the strong-coupling regime [35, 36]. In the strong-coupling regime, the coupling energy has the correlation

$$\hbar\Omega_R \propto f'^{1/2} \tag{5.2}$$

where f', is effective oscillator strength, as $f' = nf$. Moreover, n is the number of oscillators in the photon field, and f is the oscillator strength of a single molecule. Thus, the observed splitting energy is linearly correlated to the square root of integrated extinction of the system in the strong-coupling regime. This relation has been observed in the system of organic molecules in subwavelength hole arrays [7] and dyes in microcavities [4, 37, 38]. In the present system, the effective oscillator strength should increase as the number of molecules in the gap increases by repeated depositions of the dye solution. The observed linear relation in Fig. 5.3b reflects the increment of the number of dye molecules where the direction of the induced dipole matches that of the localised EM field at the gap. The observed maximum splitting energy achieved was up to 0.39 eV, which is comparable to the largest values that have been reported to date [14–16, 29]. In the present system, monomer HITC molecules may contribute to the strong coupling because the localised plasmons (730 nm, 1.70 eV) are energetically matched to the monomer absorption band (736 nm, 1.68 eV), as the H-aggregated molecules exhibit a blue-shifted absorption band (678 nm, 1.83 eV) [15].

Figure 5.4a shows the extinction spectra of the bare Ag-NSL substrates with distinct gap distances of the Ag nanodimer. The LSPR wavelength can be varied

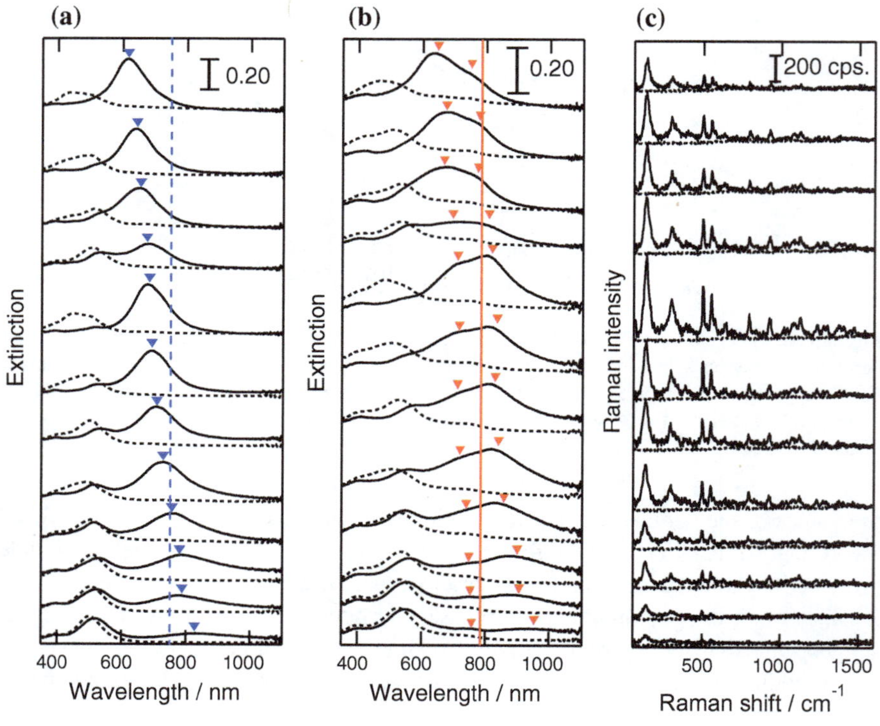

Fig. 5.4 **a** Polarised extinction spectra from the bare Ag-NSL substrates with a distinct gap distance [where the polarisation angles of incident light to the long axis of the dimers are parallel (*solid line*) and perpendicular (*dotted line*)] and the HITC extinction peak [736 nm (1.68 eV, *blue broken line*)], **b** the corresponding polarised extinction spectra of Ag@HITC, and the Raman excitation energy [785 nm (1.58 eV, *red solid line*)], **c** corresponding polarised Raman spectra observed in the Ag@HITC [scattering parallel (*solid line*) and perpendicular (*dashed line*) with extinction parallel to the dimer axis] (where I_{ex} = 50 μW, and t_{sec} = 1 s)

from 600 nm (2.07 eV) to 850 nm (1.46 eV) depending on the dimer gap distance [30]. After the substrates deposited HITC, the bare plasmon peak split into two, as shown in Fig. 5.4b. The substrate-dependent energy splits shown in Fig. 5.4b reflect the difference in the coupling strength between the dye excitons and the localised plasmons with distinct energy on their respective Ag-NSL substrates. The extinction under the perpendicular polarisation to the long dimer axis of the metal nanostructure (dotted lines in Fig. 5.4a) shows negligibly small extinction at the wavelength region of the plasmon band observed in the parallel polarisation (solid lines in Fig. 5.4a), indicating that the contribution of the extinction due to the monomer and aggregates that do not have significant coupling with the localised plasmons is relatively small (less than 2×10^{-2}).

The amounts of dye deposited on the substrates were fixed by adjusting the amount of the dye solution at the deposition to make it the same (total 600 μl). I optimised the amount of dye on the substrate, resulting in the maximum energy of Rabi splitting and avoiding the deposition of excess dye molecules that would not contribute the strong coupling. Thus, respective substrates with distinct splitting energies are expected to have comparable amounts of dye at the gap. The observed substrate dependence proves that only the excitons in the dye molecules at the gap of the metal nanostructure contribute to the apparent change in the extinction spectra shown in Fig. 5.4a and b, even though the number of molecules at the gap is much smaller than the total number of dye molecules on the substrate.

These Ag@HITS substrates showed distinct Raman intensities (Fig. 5.4c). The maximum enhancement occurred at the substrate with the maximum absorption of the polariton at the Raman excitation wavelength of 785 nm (1.58 eV; fifth from the top in Fig. 5.4c). The dependence of the Raman intensity on the substrate clearly indicates that the new hybridised states that were formed via strong coupling can contribute to the enhancement of the Raman scattering process. It should be noted that the substrates showed highly polarised Raman scattering characteristics. Apparent scattering was only observed in the parallel scattering configuration under the excitation parallel to the long axis of the dimer (solid lines in Fig. 5.4c). Raman scattering was not observed at the perpendicular scattering configuration under parallel excitation (dotted lines in Fig. 5.4c). The observed characteristics of polarised Raman measurement demonstrated that the dye molecules at the gap of the dimer contributed to the Raman scattering process. Considering that the amounts of dye molecules at the gap were comparable amongst the respective substrates, variation of the Raman scattering intensity observed in the present systems could reflect a difference in the resonance Raman process related to the new hybridised states produced by the strong coupling of the excitons with localised plasmons.

Although the intensities depend on the coupling strength, the Raman spectra were not significantly changed by the strength. The peaks of the Raman spectra at 133 and 298 cm^{-1} were assigned to the molecule's structural vibrations, while those at 501, 554, 797, 929 and 1126 cm^{-1} were assigned to the C–C stretching mode [39]. Although the Raman frequencies of these vibrational modes observed at crystalline HITC and Ag@HITC were the same, the SERS spectra of Ag@HITC

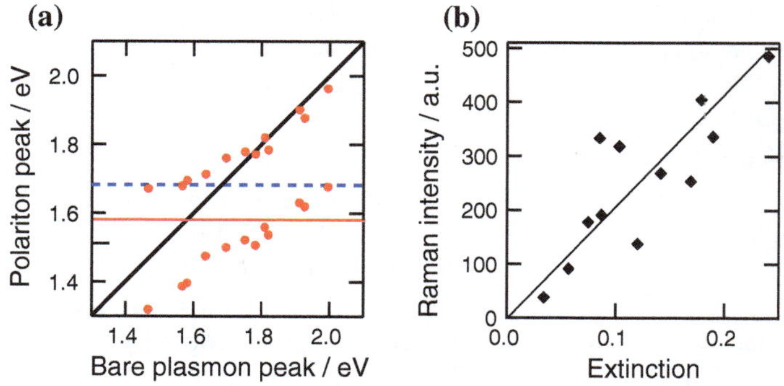

Fig. 5.5 **a** Positions of the peaks in Fig. 2b plotted against the corresponding bare LSPR positions (*red circles*), where the blue dashed line represents the uncoupled HITC exciton energy [1.68 eV (736 nm)], and the *red solid lines* represent the Raman excitation energy [1.58 eV (785 nm)]; **b** Raman intensity of 133 cm^{-1} band plotted against the corresponding extinction of Ag@HITC

showed a different relative intensity of these vibrational modes compared with those in the Raman spectra of crystalline HITC [39]. The C–C stretching modes of the SERS spectra became weaker than those of the Raman spectrum of crystalline HITC. These intensity ratios of the Raman spectra of the strong-coupling system reflect the orientations of molecules in the highly localised anisotropic EM fields at the gap of the dimer.

Figure 5.5a depicts the energy of the polariton peak against the energy of the plasmon peak observed on the bare Ag-NSL substrates. The diagonal solid line and the transverse dashed line indicate the plasmon and exciton positions, respectively. The plot shows anti-crossing behaviour, which also indicates that the present system can undergo strong coupling with LSPR and the dye excitons, as proven by the linear dependence of the variation of the Rabi splitting as a function of the square root of the integrated extinction (Fig. 5.3b). When these polariton energies are compared with the energy of the Raman excitation (785 nm, 1.58 eV) shown in Fig. 5.5a as a red line, the energy of the lower branches of the polariton matches that of the Raman excitation in the present system, especially for the substrate with the plasmon peak at 1.8 eV. This substrate shows the most intense Raman scattering from molecules in the strong-coupling regime, as shown in the spectrum (fifth from the top in Fig. 5.4). The excitation laser wavelength corresponds to the lower energy polariton of the substrate shown at the top of Fig. 5.4b and to the higher energy polariton at the bottom of Fig. 5.4b.

To estimate the effect of the excitation of the upper and the lower branches of the polariton, the observed Raman intensity was compared with the extinction intensity of the polariton. Figure 5.5b shows the Raman intensity in the 133 cm^{-1} band against the corresponding extinction intensity at Ag@HITC. The Raman intensity increases linearly with the increasing the extinction of the polariton. This linear relationship suggests that strong enhancement of the Raman process can occur when a polariton excitation lies within the Raman excitation [1.58 eV (785 nm)] and scattering processes [133 cm^{-1} Raman-shifted photons at 1.56 eV (793 nm)]. A strong Raman signal originates from the resonance with the upper and lower branches of the polariton. In the case of cavity Raman enhancement in a semiconductor quantum well, the polariton in the lower branch shows higher enhancement than that in the upper branch [40]. The slower dephasing of the polaritons in the lower branch has been considered to result in higher Raman enhancement. In the strong-coupling regime between the dye exciton and LSP, photoluminescence is only observed from the lower polariton branch, because the dephasing rate of the polariton in the higher polariton branch is faster than that in the lower polariton branch [6]. Further quantitative measurement using a system with a controlled number of molecules at the nanogap could clarify the difference in the effect of the excitation to the upper and the lower branches of the polariton on the Raman scattering from molecules under strong-coupling conditions at LSP. Raman scattering intensity may also reflect the effects of the coupling of the light in and out with plasmonic metal nanostructures. The Purcell effect could be one possible way to change the scattering efficiency of confined light far below the diffraction limit. At the present, the contribution of Purcell effect has not yet been clarified, while previously documented results have been shown in SERS [41] and other fluorescence enhancement experiments [42].

5.4 Conclusion

In conclusion, I prepared a system that showed strong coupling between the LSPR and dye excitons using Ag dimer structures and HITC molecules. Despite the broad absorption band of HITC, a relatively large Rabi splitting energy could be achieved by tuning the LSPR energy through careful control of the metal nanogap distance. Polarised Raman measurement proved the contribution to Raman enhancement of the strong coupling at the nanogap, indicating that well-defined interactions occur between the molecule electronic states and the confined EM field. The maximum Raman enhancement was obtained at the resonant energy between the hybrid states and the excitation. The Raman enhancement in the strong-coupling regime allowed us to observe the vibrational structure in the new hybridised molecule states for the first time. This will guide us towards a detailed understanding of the excitation process in the selectively polarised excitation of molecules in confined nanosize EM fields. The system described here offers interesting possibilities for use of the strong interaction between the molecules and the localised plasmons, and may possibly be applied to the development of a novel photochemical reaction using LSPs.

References

1. K.H. Drexhage, IV Interaction of light with monomolecular dye layers, in *Progress in Optics*, vol 12, ed. by E. Wolf (Elsevier, 1974), pp. 163–232
2. B. Pettinger, J. Chem. Phys. **85**, 7442–7451 (1986)
3. I. Pockrand, A. Brillante, D. Mobius, J. Chem. Phys. **77**, 6289–6295 (1982)
4. D.G. Lidzey, D.D.C. Bradley, M.S. Skolnick, T. Virgili, S. Walker, D.M. Whittaker, Nature **395**, 53–55 (1998)
5. D.G. Lidzey, Science **288**, 1620–1623 (2000)
6. J. Bellessa, C. Bonnand, J. Plenet, J. Mugnier, Phys. Rev. Lett. **93**, 036404 (2004)
7. J. Dintinger, S. Klein, F. Bustos, W. Barnes, T. Ebbesen, Phys. Rev. B **71**, 035424 (2005)
8. Y. Sugawara, T.A. Kelf, J.J. Baumberg, M.E. Abdelsalam, P.N. Bartlett, Phys. Rev. Lett. **97**, 266808 (2006)
9. T. Hakala, J. Toppari, A. Kuzyk, M. Pettersson, H. Tikkanen, H. Kunttu, P. Törmä, Phys. Rev. Lett. **103**, 053602 (2009)
10. J.A. Hutchison, T. Schwartz, C. Genet, E. Devaux, T.W. Ebbesen, Angew. Chem. Int. Ed. **51**, 1592–1596 (2012)
11. J.A. Hutchison, A. Liscio, T. Schwartz, A. Canaguier-Durand, C. Genet, V. Palermo, P. Samori, T.W. Ebbesen, Adv. Mater. **25**, 2481–2485 (2013)
12. G.P. Wiederrecht, G.A. Wurtz, J. Hranisavljevic, Nano Lett. **4**, 2121–2125 (2004)
13. G.A. Wurtz, P.R. Evans, W. Hendren, R. Atkinson, W. Dickson, R.J. Pollard, A.V. Zayats, W. Harrison, C. Bower, Nano Lett. **7**, 1297–1303 (2007)
14. N.T. Fofang, T.H. Park, O. Neumann, N.A. Mirin, P. Nordlander, N.J. Halas, Nano Lett. **8**, 3481–3487 (2008)
15. W. Ni, T. Ambjornsson, S.P. Apell, H. Chen, J. Wang, Nano Lett. **10**, 77–84 (2010)
16. A.E. Schlather, N. Large, A.S. Urban, P. Nordlander, N.J. Halas, Nano Lett. **13**, 3281–3286 (2013)
17. E.C. Ru Lu, P.G. Etchegoin, *Principles of Surface-Enhanced Raman Spectroscopy: and related plasmonic effects.* (Elsevier Science, 2008)
18. O.V. Makarova, A.E. Ostafin, H. Miyoshi, J.R. Norris, D. Meisel, J. Phys. Chem. B **103**, 9080–9084 (1999)
19. N. Kometani, M. Tsubonishi, T. Fujita, K. Asami, Y. Yonezawa, Langmuir **17**, 578–580 (2001)
20. S. Franzen, J.C.W. Folmer, W.R. Glomm, R. O'Neal, J. Phys. Chem. A **106**, 6533–6540 (2002)
21. B.I. Ipe, K.G. Thomas, J. Phys. Chem. B **108**, 13265–13272 (2004)
22. I.I.S. Lim, F. Goroleski, D. Mott, N. Kariuki, W. Ip, J. Luo, C.J. Zhong, J. Phys. Chem. B **110**, 6673–6682 (2006)
23. M. Moskovits, J. Raman Spectrosc. **36**, 485–496 (2005)
24. D.A. Weitz, S. Garoff, J.I. Gersten, A. Nitzan, J. Chem. Phys. **78**, 5324–5338 (1983)
25. W.H. Yang, G.C. Schatz, R.P. Vanduyne, J. Chem. Phys. **103**, 869–875 (1995)
26. J. Jiang, K. Bosnick, M. Maillard, L. Brus, J. Phys. Chem. B **107**, 9964–9972 (2003)
27. A.M. Michaels, J. Jiang, L. Brus, J. Phys. Chem. B **104**, 11965–11971 (2000)
28. A.M. Kelley, J. Chem. Phys. **128**, 224702 (2008)
29. N. Cade, T. Ritman-Meer, D. Richards, Phys. Rev. B **79**, 241404 (2009)
30. Y. Sawai, B. Takimoto, H. Nabika, K. Ajito, K. Murakoshi, J. Am. Chem. Soc. **129**, 1658–1662 (2007)
31. M. Takase, Y. Sawai, H. Nabika, K. Murakoshi, J. Photochem. Photobiol., A **221**, 169–174 (2011)
32. M. Takase, H. Nabika, S. Hoshina, M. Nara, K. Komeda, R. Shito, S. Yasuda, K. Murakoshi, Phys. Chem. Chem. Phys. **15**, 4270–4274 (2013)
33. Y. Sawai, B. Takimoto, H. Nabika, K. Ajito, K. Murakoshi, Faraday Discuss. **132**, 179–190 (2006)

34. S. Savasta, R. Saija, A. Ridolfo, O. Di Stefano, P. Denti, F. Borghese, ACS Nano **4**, 6369–6376 (2010)
35. M. Takase, H. Ajiki, Y. Mizumoto, K. Komeda, M. Nara, H. Nabika, S. Yasuda, H. Ishihara, K. Murakoshi, Nature Photon. **7**, 550–554 (2013)
36. C.M. Aikens, L.R. Madison, G.C. Schatz, Nature Photon. **7**, 508–510 (2013)
37. P.A. Hobson, W.L. Barnes, D.G. Lidzey, G.A. Gehring, D.M. Whittaker, M.S. Skolnick, S. Walker, Appl. Phys. Lett. **81**, 3519 (2002)
38. M.S. Skolnick, T.A. Fisher, D.M. Whittaker, Semicond. Sci. Technol. **13**, 645 (1998)
39. D.L. Akins, S. Özçelik, H.-R. Zhu, C. Guo, J. Phys. Chem. A **101**, 3251–3259 (1997)
40. A. Fainstein, B. Jusserand, Raman scattering in resonant cavities, in *Light Scattering in Solid IX*, vol 108, ed. by M. Cardona, R. Merlin (Springer Berlin Heidelberg, 2007), pp. 17–110
41. K.-I. Yoshida, T. Itoh, H. Tamaru, V. Biju, M. Ishikawa, Y. Ozaki, Phys. Rev. B **81**, 115406 (2010)
42. H. Yuan, S. Khatua, P. Zijlstra, M. Yorulmaz, M. Orrit, Angew. Chem. Int. Ed. **52**, 1217–1221 (2013)

Chapter 6
Electrochemical Control of Strong Coupling Between Localised Surface Plasmons and Dye Excitons

6.1 Introduction

It was shown in Chap. 5 that LSPR can be strongly coupled to the excited states of molecules situated on the nanosized metal surface. An LSP can be regarded as confined light to provide a highly polarised EM field with a specific resonant energy that can effectively interact with excitons of materials [1–9]. Thus, especially in the strong-coupling regime using LSPs, these properties have attracted wide research interest as an entirely new concept of molecular plasmonics. The excited state is considered a new energy state that is not limited by materials' inherent energy state. This concept can be used to extend plasmon-induced photoexcitation processes. If we can control the state in the strong-coupling regime, this can be used as an approach for tuning the photoexcited state. In the propagation of plasmon resonance using a flat metal surface [10, 11], nanohole array [12] and nanograting [13, 14], it was reported that the strong-coupling state modulated the lifetime of the excited states and the absorption and emission properties [15–17]. The modulated property was reflected in the nonlinear reactivity of the chemical reaction [18] and by tuning the work function [19] via strong coupling and thermodynamics [20].

Propagating SPR can be easily used to evaluate whether the system has achieved a strong-coupling regime. The energy dispersion was evaluated by changing the angle of the incident light, which caused the wavevector to change in about the same environment as the substrate. In addition, the photoisomerisation of dye molecules can tune the effective oscillator strength by simply light illuminating or heating the substrate [21, 22] and gas flow [23]. In contrast to the propagating SPR, the energy of the LSPR was defined individually by the shape and size of the nanostructure. Distribution of the plasmon energy led to differences in the state of the strong coupling. At present, control in the coupling of the system can be only achieved by tuning the properties of dye molecules.

F. Nagasawa, *Studies on the Plasmon-Induced Photoexcitation Processes of Molecules on Metal Surfaces*, Springer Theses,
DOI 10.1007/978-4-431-56579-6_6

It is known that the resonant energy of LSPR can achieve control by tuning the electronic property in the electrochemical environment. In the metal nanoparticle system, it has been found that the LSP band moves lower (or higher) as the core is electrolytically made more electron deficient (or excessive) [24, 25]. Electrochemical potential tuning in the strong-coupling regime can characterise the LSP energy, which illustrates the dispersion of the coupling. In this chapter, electrochemical potential control is adopted for tuning the LSPR energy and the definition of the electronic properties of the strong-coupling regime. Raman spectra are also acquired to explore the molecular state in the strong-coupling regime.

6.2 Experimental

An Au nanostructure (Au-NSL) was prepared using the angle-resolved nanosphere lithography technique. For electrochemical measurement, a PS bead (diameter ϕ = 350 nm) monolayer was prepared on ITO glass. Au monomer structures were prepared after single deposition with a vertical angle on the glass and thicknesses of 20, 30 and 40 nm. Following this, the PS mask was removed via sonication in Milli-Q water. The dye used was HITC. After dye deposition (600 μL/cm^{-1}), residual dye molecules were rinsed in Milli-Q water. Dye-coated Au monomer on ITO glass and bare Au monomer on ITO glass were used as working electrodes. The Pt plate was used as the counter electrode. The potential of the working electrode was regulated with respect to the Ag/AgCl reference electrode.

In the electrochemical environment, the extinction spectrum and the Raman spectrum were acquired in situ. A three-electrode electrochemical cell was designed. The extinction spectrum of the Ag dimer array in the visible-NIR region was recorded utilising a multichannel spectrometer (MCPD-2000, Ohtsuka Electronics; observed area was ~0.3 cm^2).

In the electrochemical environment, Raman measurement was performed in the backscattering configuration. An objective lens (×100) immersed in water was used for in situ Raman measurement. The excitation energy was λ_{ex} = 785 nm, and the exposure time was t = 1 s. The expanded NIR beam was focussed onto the sample using a water immersion objective lens with ×100 magnification and a numerical aperture of 1.0. The estimated spot size of irradiation was approximately 1 μm, and the output intensity was 50 μW. All Raman measurements were carried out in situ by immersion of the Au-NSL substrate into aqueous solution containing 0.1 M $NaClO_4$ as the electrolyte.

The structure of the Au monomer on the ITO glass substrate was inspected using an AFM (Nanoscope-IIIa, Digital Instruments) in air and a scanning electron microscope (SEM, JSM-6700FT, JEOL).

6.3 Results and Discussion

The Au-NSL structures on ITO glass substrate are shown in Fig. 6.1. Nanostructures were prepared by changing the Au deposition thickness (20, 30, 40 nm) on the ITO glass. Figure 6.1a, d, g shows deposition at 20 nm (Au20), Fig. 6.2b, e, h shows deposition at 30 nm (Au30) and Fig. 6.1c, f, i shows deposition of 40 nm (Au40). Triangle monomers were observed from SEM imaging (Fig. 6.1 top and middle) and the height was confirmed from AFM imaging (Fig. 6.1 bottom). The length of the perpendicular bisectors of the equilateral Au triangles was approximately 90 nm.

The extinction spectra of Au-NSL immersed in 0.1 M $NaClO_4$ aqueous solution are exhibited in Fig. 6.2a. These structures showed peaks around 800 nm, and the peak energy shifted depending on the thickness of Au. The maximum peak wavelength was plotted against the Au nanoparticles' height, as shown in Fig. 6.2b. The higher nanoparticle showed the blue-shifted LSP peak. The observed peak in 0.1 M $NaClO_4$ aqueous solution was blue shifted compared with that in air. The

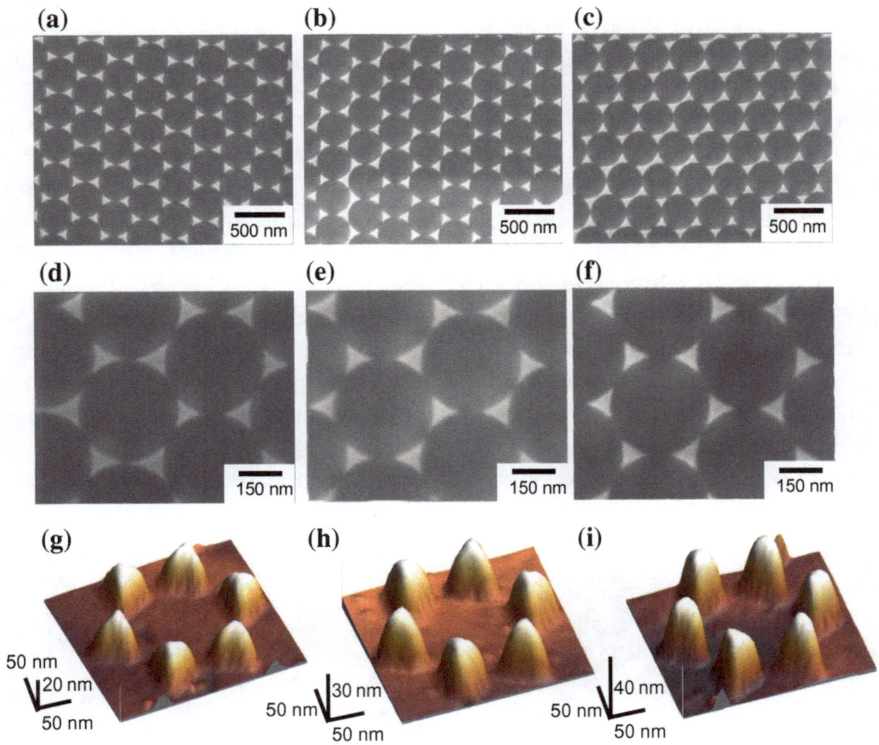

Fig. 6.1 **a–f** SEM and **g–i** AFM image of Au-NSL with Au thicknesses of 20 nm (*a*, *d*, *g* Au20), 30 nm (*b*, *e*, *h* Au30) and 40 nm (*c*, *f*, *i* Au40)

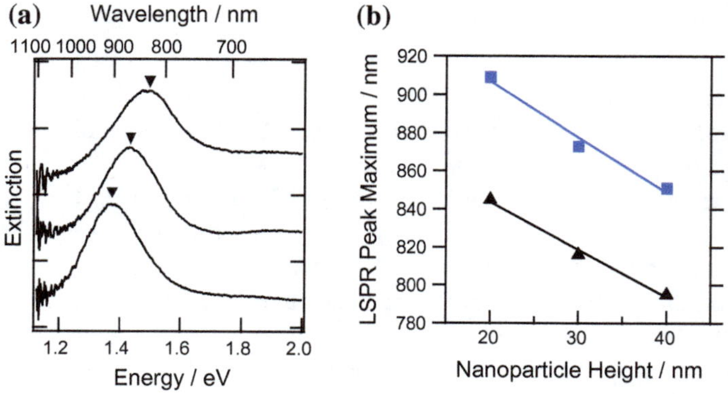

Fig. 6.2 **a** Extinction spectra of Au-NSL and **b** LSP peak maximum shift against nanoparticle height: Au height 20 nm (**a**, *top*), 30 nm (**a**, *middle*), 40 nm (**a**, *bottom*); LSPR peak in air (**b**, *square*) in 0.1 M $NaClO_4$ aqueous solution (**b**, *triangle*)

peak shift occurred because of the change in the dielectric environments in air ($n = 1.00$) in relation to 0.1 M $NaClO_4$ aqueous solution ($n = 1.33$) [26].

Figure 6.3 shows the extinction spectrum under the electrochemical potential control of the Au-NSL substrate deposited with HITC (top: Au@HITC) and the bare Au-NSL substrate (bottom). After deposition of the HITC, the observed peaks at +0.3 V were 1.48 eV (839 nm) in Au20@HITC (Fig. 6.3a), and the peak was shifted to 1.58 eV (787 nm) at −0.9 V. In Au30@HITC (Fig. 6.3b) the peaks at +0.3 V were 1.69 eV (735 nm) and 1.50 eV (826 nm), and they were blue shifted to 1.76 eV (705 nm) and 1.65 eV (752 nm) at –0.9 V. In Au40@HITC (Fig. 6.3c), the peak at +0.3 V was 1.79 eV (693 nm) and blue shifted to 1.79 eV (692 nm) at –0.9 V, while 1.64 eV (758 nm) at +0.3 V showed decreased intensity.

In the bare Au substrate, the observed peaks of Au20 were 1.36 eV (909 nm) at +0.3 V and 1.45 eV (855 nm) at –0.9 V (Fig. 6.3d), the observed peaks of Au30 were 1.42 eV (873 nm) at +0.3 V and 1.52 eV (814 nm) at –0.9 V (Fig. 6.3e) and the observed peaks of Au40 were 1.46 eV (851 nm) at +0.3 V and 1.55 eV (802 nm) at –0.9 V (Fig. 6.3f). The behaviour of the blue shift of the LSPR energy under positive polarisation was reported previously [24, 25]. In the bulk metal, the real part of the dielectric function can be expressed approximately as

$$\varepsilon_c' = \varepsilon^{\infty} - \frac{\lambda_{\mathrm{peak}}^2}{\lambda_{\mathrm{p}}^2}, \tag{6.1}$$

where λ_{p} is the bulk metal plasmon wavelength. λ_{p} is given by

$$\lambda_{\mathrm{p}} = \frac{2\pi c}{\omega_{\mathrm{p}}}, \tag{6.2}$$

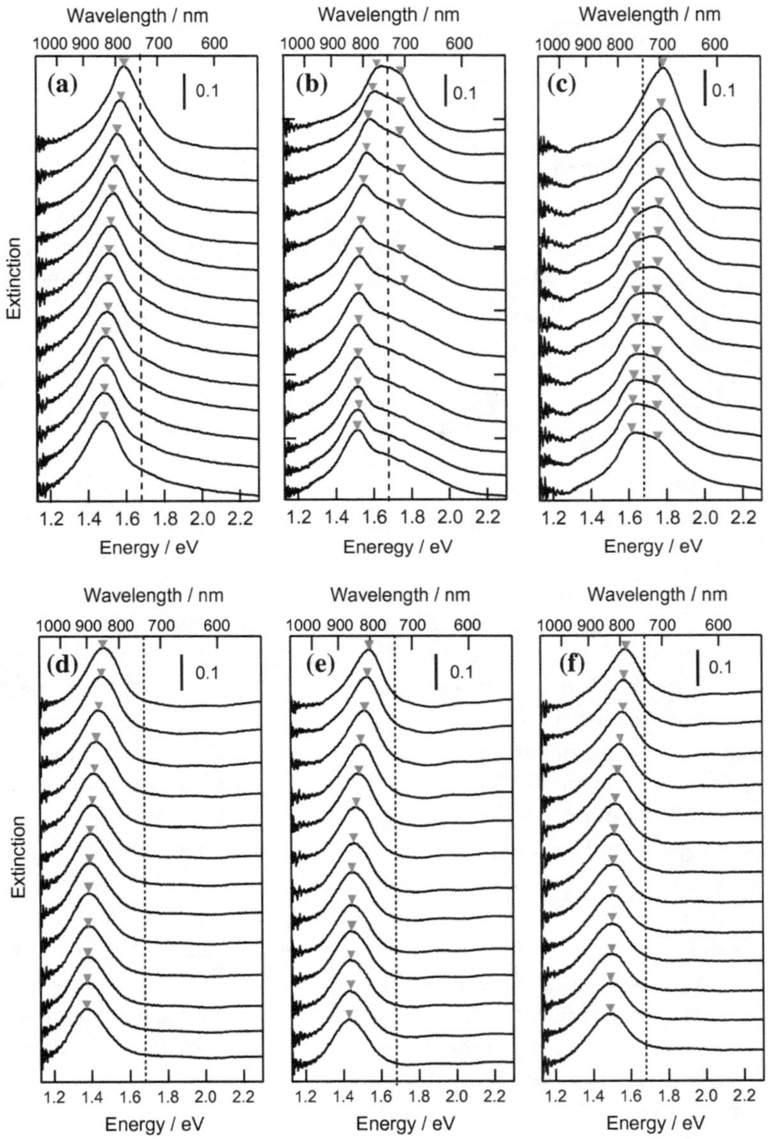

Fig. 6.3 Extinction spectra in 0.1 M $NaClO_4$ aqueous solution: **a** Au20@HITC, **b** Au30@HITC, **c** Au40@HITC, **d** Au20, **e** Au30, **f** Au40. The dotted line is the exciton energy of HITC. The electrochemical potentials were from +0.3 V (*bottom*) to –0.9 V (*top*); spectra were acquired at each 0.1 V change

where c is the speed of light in vacuum and ω_p is the metal's bulk frequency. ω_p is given by

$$\omega_p = \left(\frac{Ne^2}{m\varepsilon_0}\right)^{1/2}, \tag{6.3}$$

where N is the free electron concentration and m is the electron mass. In the electrochemical condition, N can be varied by changing electrochemical potential of metals, leading to a change in ω_p. The relationship between the free electron concentration and the SP band wavelength is derived from Eqs. (6.1), (6.2) and (6.3) to give

$$\left[\frac{\lambda_{peak}^{final}}{\lambda_{peak}^{init}}\right] = \left[\frac{N_{init}}{N_{final}}\right]^{1/2}, \tag{6.4}$$

where λ_{peak}^{init}, λ_{peak}^{final} are the peak wavelengths in initial and final state, respectively, and N_{init}, N_{final} are the free electron concentrations. This relation can be applied to LSPR for metal nanoparticles. In the present system, negative polarisation from +0.3 to –0.9 V resulted in the blue shift of λ_{peak} by 54 nm for Au20, reflecting an estimated decrease in N of 3.1%. At Au30 and Au40, the shift of 59 nm and 49 nm corresponds to estimated decreases in N of 3.5% and 3.0%, respectively. A comparable change in λ_{peak} was reported previously in a system of Au nanoparticles on Pt mesh substrate in 0.1 M Bu_4NPF_6/CH_2Cl_2. Positive polarisation of about 1.0 V from –0.16 to +0.82 V versus Ag/Ag^+ resulted in a 9-nm shift of λ_{peak} from 525 nm to 516 nm, corresponding to a 5% decrease in N. Based on these observations in the present system, I was able to estimate the possible shift of λ_{peak} depending on the electrochemical potential of the system in the strong-coupling regime. I assumed that the peak shift of λ_{peak} in the dye-deposited substrate should be comparable to that without dye because the thickness is same.

At –0.9 V in Au30@HITC (Fig. 6.3b), the centre of the peak energy is greater than the energy of HITC excitons. When the dye exciton energy and plasmon energy were the same, strong coupling was achieved; in addition, the energy separation was the same due to the exciton energy. Thus, the plasmon mode at –0.9 V was estimated to have the same energy as the HITC exciton (1.68 eV). In +0.3 V of Au30@HITC, the LSPR position was estimated from the ratio of free electron concentration to Au30 using Eq. (6.4). Figure 6.4 shows the image plot of the extinction spectra of Au@HITC at each Au monomer height. The estimated LSP position (Fig. 6.4, white circle) and exciton energy position of HITC (Fig. 6.4, white square) were plotted against the electrochemical potential. As shown in Fig. 6.4c, at +0.3 V, the LSPR and HITC energy were coincident and the strong-coupling regime was achieved; at –0.9 V, the LSPR did not coincide with the HITC exciton energy. Thus, electrochemical potential control can modulate the LSPR energy, which induces the strong-coupling region.

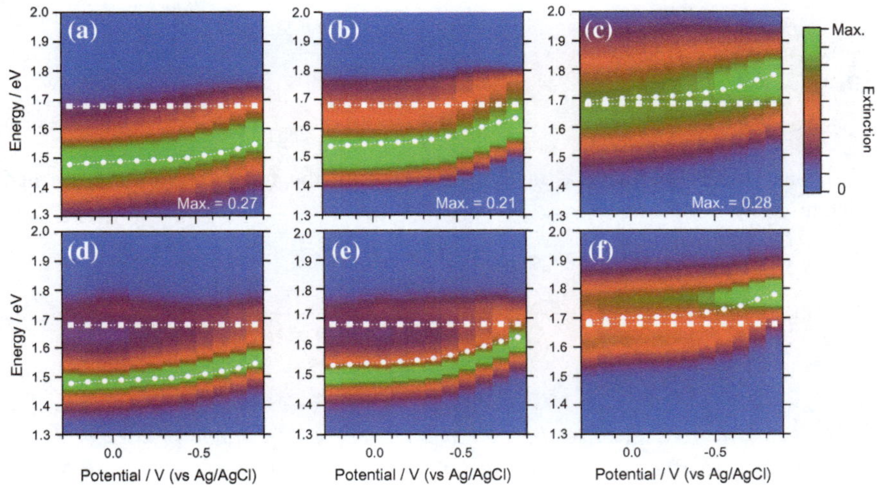

Fig. 6.4 Image plot of the extinction spectra against the electrochemical potential: **a** Au20@HITC, **b** Au30@HITC, **c** Au40@HITC. The exciton energy of HITC (1.68 eV, *circle*) and the estimated LSPR band energy (*square*) were plotted against the electrochemical potential

In cavity quantum electronic dynamics, the coupling between the plasmonic structure and excitons of dye molecules or quantum dots has been evaluated using the classical mechanical model [27]. These results showed good recurrence of the finite difference time domain calculations [27]. I adopted the classical coupled oscillator model to analyse the spectral features. Both oscillators were coupled through the electric near field with coupling energy $\hbar g = |\langle \mathbf{d} \cdot \mathbf{E} \rangle|$, where **d** represents an electric dipole due to the electronic transition resonance of a molecule and **E** represents an electric dipole due to plasmonic resonance at the enhanced field of the metal surface. The equations of motion for the two oscillators are

$$\frac{\partial^2 x_{\mathrm{LSP}}(t)}{\partial t^2} + \gamma_{\mathrm{LSP}} \frac{\partial x_{\mathrm{LSP}}(t)}{\partial t} + \omega_{\mathrm{LSP}}^2 x_{\mathrm{LSP}}(t) + g \frac{\partial x_{\mathrm{HITC}}(t)}{\partial t} = F_{\mathrm{LSP}}(t), \tag{6.5}$$

$$\frac{\partial^2 x_{\mathrm{HITC}}(t)}{\partial t^2} + \gamma_{\mathrm{HITC}} \frac{\partial x_{\mathrm{HITC}}(t)}{\partial t} + \omega_{\mathrm{HITC}}^2 x_{\mathrm{HITC}}(t) - g \frac{\partial x_{\mathrm{LSP}}(t)}{\partial t} = F_{\mathrm{HITC}}(t), \tag{6.6}$$

where x_{LSP} and x_{HITC} are the coordinates of plasmonic and HITC electronic oscillation, respectively; γ_{LSP} and γ_{HITC} are the line widths of plasmonic and HITC electronic resonance, respectively; ω_{LSP} and ω_{HITC} are the resonance frequencies of plasmons and the HITC electronic transition, respectively; and F_{LSP} and F_{HITC} represent the driving forces of the two oscillators. I set $F_{\mathrm{HITC}}(t) = 0$ because the molecular extinction cross section is negligible compared to that of the Au nanoparticle, $F_{\mathrm{LSP}} \ll F_{\mathrm{HITC}}(t)$. I can describe the extinction cross section as

$$C_{ext}(\omega) \propto \omega \,\mathrm{Im}\left(\frac{\left(\omega_{HITC}^2 - \omega^2 - i\gamma_{HITC}\omega\right)}{\left(\omega^2 - \omega_{LSP}^2 + i\gamma_{LSP}\omega\right)\left(\omega^2 - \omega_{HITC}^2 + i\gamma_{HITC}\omega\right) - \omega^2 g^2}\right). \tag{6.7}$$

To make clear the physical insight of Eq. (6.7), the formula can be deformed under resonant condition $\omega_0 = \omega_{LSP} = \omega_{HITC}$ as

$$C_{ext}(\omega) \propto \omega \mathrm{Im}\left(\frac{\omega_0 - \omega - i\gamma_{HITC}\omega/2}{2\omega_0\sqrt{g^2 - (\gamma_{LSP} - \gamma_{HITC})^2/4}}\left(\frac{1}{\omega - \Omega_+} - \frac{1}{\omega - \Omega_-}\right)\right) \tag{6.8}$$

where $\Omega_\pm = \omega_0 + i(\gamma_{LSP} + \gamma_{HITC})/4 \pm \sqrt{g^2/4 - (\gamma_{LSP} - \gamma_{HITC})^2/16}$. I simulated the spectral properties in Fig. 6.4 using the peak position of LSP energy and linewidth and HITC energy and linewidth. The peak positions of experimental extinction spectra were fitted by the Lorentzian function.

Simulated spectra are shown in Fig. 6.4d, e, f, exhibiting good reappearance. The spectral features were simulated by changing the value of g. In doing this, I was able to obtain the coupling strength g, as shown in Fig. 6.5. From +0.3 to –0.5 V, the values of g were constant at around 0.15–0.20 eV; the g values then diminished in the area from –0.5 to –0.9 V. The value of g (c) is determined by the electric dipole due to the electronic transition resonance of the molecule and the electric dipole resulting from plasmonic resonance in the enhanced field of the metal surface. In my system, the electric dipole was assumed to be constant for each electrochemical potential region because intrinsic Au nanostructures are used. The

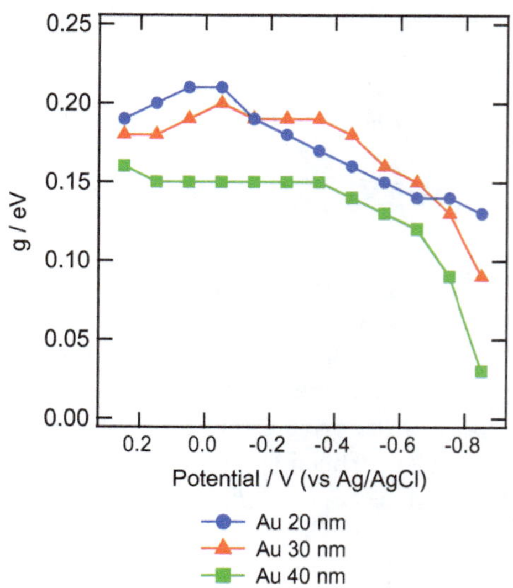

Fig. 6.5 Values of g against the electrochemical potential

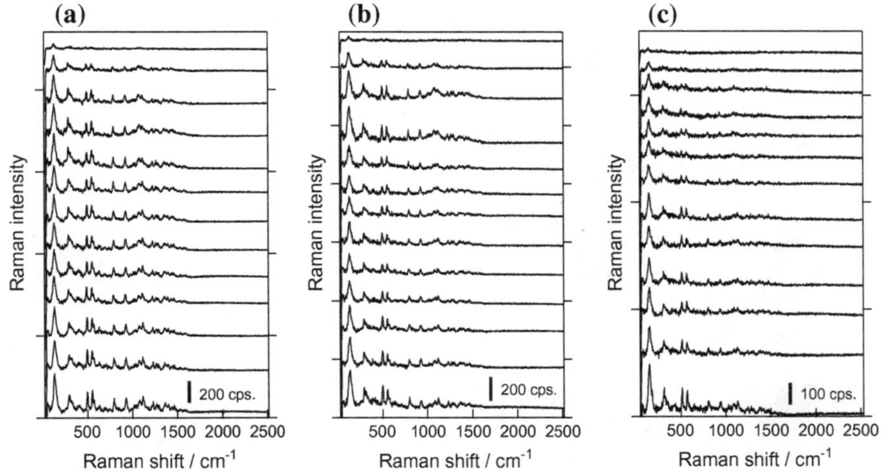

Fig. 6.6 Raman spectra under electrochemical potential control: **a** Au20@HITC, **b** Au30@HITC, **c** Au40@HITC; excitation power I_{ex} = 50 μW, exposure time t_{sec} = 1 s

diminished value of g in the negative potential region (from –0.5 to –0.9 V) may reflect the change of the electric dipole due to the electronic transition resonance of a molecule. It is known that the reduction potential of HITC molecules is –0.5 V (vs. Ag/AgCl) [28]. Fluctuations of orientation and adsorbed structure of molecules on the metal surface caused diminished value of g. In the electrochemical potential region from +0.3 to –0.5 V, I succeeded in tuning the LSP energy against the energy of HITC excitons. To evaluate the molecular fluctuation, SERS is a desirable measurement tool. After using SERS, in situ electrochemical Raman measurement was performed.

Figure 6.6 shows the Raman spectra under electrochemical potential control. The observed Raman bands were 133 and 298 cm^{-1}, which are assigned to the molecule's structural vibrations, and 501, 554, 797, 929 and 1126 cm^{-1}, which were assigned to the C–C stretching mode. The modes were the same as for HITC powder. The vibrational modes of 133 cm^{-1} provided intensity modulation for each electrochemical potential.

At the negative potential region, the Raman spectra fluctuated. Spectra fluctuating at –0.6 V are shown in Fig. 6.7a. The background of the Raman spectrum changed in particular, although the band intensity at 133 cm^{-1} was stable. Then Raman spectra were divided into the intensity of the Raman band at 133 cm^{-1} (I_{133}) and the intensity of the background (I_{BG}). In Fig. 6.7b, the Raman intensity at 133 cm^{-1} and background intensity of Au30@HITC are plotted as a function of time. The scan from positive potential to negative potential promoted the fluctuation of the Raman intensity. The reduction potential of HITC molecules is known to be –0.5 V (vs. Ag/AgCl) [28]. At the more negative reduction potential region, the

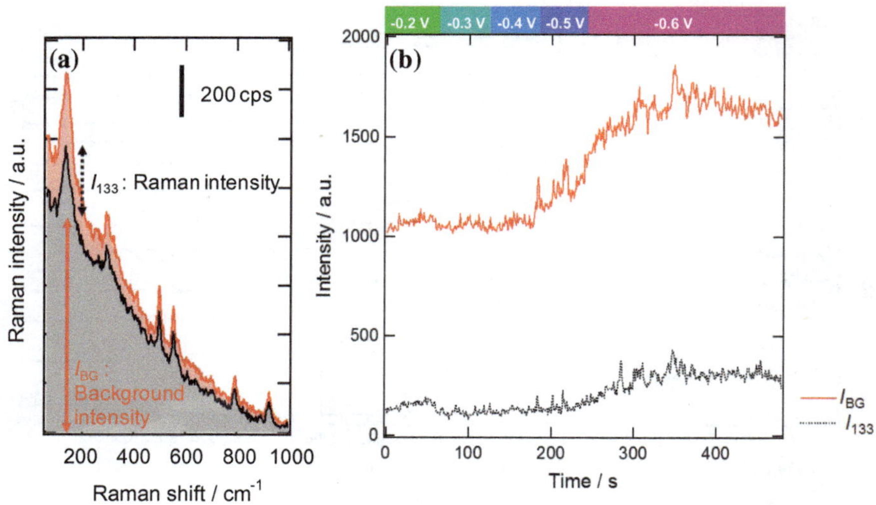

Fig. 6.7 Raman spectra at −0.6 V at 300 s (**a**, *black spectrum*) and 330 s (**a**, *red spectrum*), time course of Raman intensity (**b**, *red solid line*) and backscattering intensity (**b**, *black dotted line*)

adsorbed molecules on the metal surface seemed to diffuse. In the electrochemical tuning of energy at LSP resonance, fluctuations of the band intensity were measured. As shown in the Raman spectra, the background showed the most fluctuation. As illustrated in Fig. 6.8, background intensity and Raman intensity were analysed for each substrate. In Fig. 6.8a, b, the intensity of the backgrounds fluctuates more than the Raman intensity. Particularly in Fig. 6.8b, the fluctuation of the background intensity was highest in the Au30 substrate. At –0.2 V, which is more positive than the reduction potential, fluctuations of the intensity of the Raman band and the background intensity were both repressed, as shown in Fig. 6.8c, d. Au30 was resonant with the excitons of dye molecules in the negative potential region, which achieved a strong-coupling regime. The resonant condition induced the enhancement of the background.

For a molecular optical transition that is resonant with excitation energy, SERRS is usually accompanied by surface-enhanced fluorescence (SEF), as represented in the background. SERRS and SEF were both considered in relation to the EM effect caused by plasmons [29–33]. The blinking and fluctuation of SERRS and SEF features were also measured [34–36]. It is known that the origin of the fluctuation of the background involves changes in the orientation of dipole axis of molecules to the localised EM field and/or the distance between molecules and the metal surface [37–39]. In Fig. 6.8, increments of Raman and background intensity occurred in a time scale of several tens of seconds, which is likely to correspond to molecular diffusive motions changing their orientation and the distance from the metal surface.

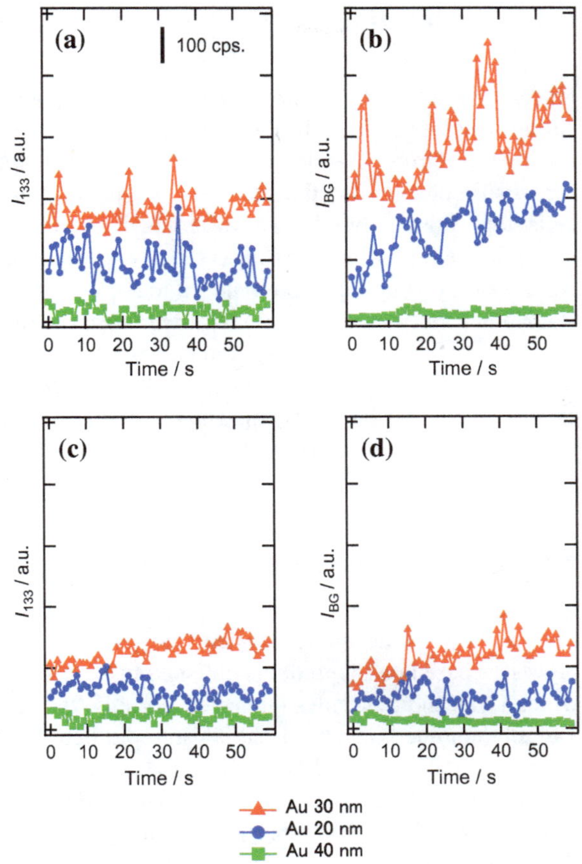

Fig. 6.8 **a**, **c** Time course of Raman intensity and **b**, **d** backscattering intensity at **c**, **d** –0.2 V and **a**, **b** –0.6 V. The Au deposition thicknesses were 20 nm (*blue circle*), 30 nm (*red triangle*) and 40 nm (*green square*)

In Fig. 6.8b, the intensity of background at Au30 especially showed larger fluctuations than the other substrates. In the strong-coupling regime, perturbation by photons to molecules is strongly confined to the molecular scale [7] because of increments in absorption intensity. The enhanced absorption cross section could induce the effect of the optical trapping of molecules near the metal nanodimer. As a consequence, moving of the molecular position caused a change in the coupling strength g ($\hbar g = |\langle \mathbf{d} \cdot \mathbf{E} \rangle|$), which was related to the molecular orientation. The background, which is reflected by the extinction spectral shape of the strong-coupling state, seems sensitive to the molecular position and orientation. Although other effects due to heat, photobleaching, photoinduced CT between metal and molecules and so on may also contribute to background fluctuation, the apparent effect of the electrochemical potential in the present system strongly supports the importance of strong coupling tuned by LSP energy to manipulate molecules in the vicinity of metal nanostructures.

6.4 Conclusion

Under the control of the electrochemical potential of the system in the strong-coupling regime, LSPR band energy was tuned by changing the density of the free electrons. While the surrounded environment was maintained by the adsorbed molecules on the metal surface under potential control, strong coupling behaviour was analysed. The switching between a weak-coupling regime and a strong-coupling regime was observed. This technique realised the evaluation and control of the coupling behaviour without changing of coverage of dye molecules. Raman enhancement was also measured when the resonance between the polariton states by strong coupling and Raman excitation energy was satisfied. This represents the first technique for tuning the energy between the LSPR and molecular exciton dye in the same system. The possibility of tuning the coupling strength can provide insight into the control of the photoexcitation state.

6.5 General Conclusion

In this thesis, I developed systems to pave the way for novel photoexcitation processes based on the findings obtained from in situ observation of the systems, in which LSPs and molecules interacted strongly. It was found that the use of LSPs, as a highly confined, localised and anisotropic light source, was appropriate and relevant for evaluating and controlling the interaction. Highly ordered metal nanostructures providing the possibility of anisotropic photoexcitation at the interface were suggested. The optimisation of the system, which was confirmed from the singularities of analyte, evaluation of the structures and control of the electrochemical potential of the metal surface, was carried out through the anisotropic excitation from LSP and an additional transition for the photoexcitation process. Furthermore, electronic energy excitation states were modulated by the strong coupling between LSP and excitons of molecules. This developed feature of the electronic structure was actively controlled by the electrochemical potential. The construction of the developed electronic structure systems paves the way for the plasmon-induced photoexcitation processes. The main results are summarised below.

Initially, an Ag nanodimer structure was fabricated for polarised SERS measurement (Chap. 2). The polarised SERS were performed in the 4,4'-bipyridine (44bpy) solution. Depolarised SERS signals were obtained at the specific band intensity ratio of the Raman spectrum. I evaluated the single crystal 44bpy structure and the Raman tensor, which indicated the Raman activity. By comparing the results of DFT calculation, it was found that the Raman intensity ratio represented the adsorbed molecular orientation on the metal surface. The SERS depolarisation at specific adsorption states of molecules implied that the interface between molecules and metal induces anisotropic photoexcitation. For evaluation of the photoexcitation process, I performed single molecular measurement (Chap. 3) and measurement under electrochemical potential tuning (Chap. 4).

The single-molecule Raman measurement was achieved by simultaneous conductance measurement. The metal nanojunction was prepared using the electron beam lithography and sputtering methods. The conductance between the metal nanojunction clarified the number of molecules and the strength of adsorption. The CT from metal to molecules provided an additional electronic resonant condition in the excitation energy. The resonant-like condition of the Raman process showed the enhancement of an additional vibrational mode that was not observed in the normal Raman condition. This Raman process apparently showed the anisotropic excitation at the interface between molecules and metal. The system was then developed for SWNT measurement under electrochemical potential tuning. SWNTs are appropriate for evaluation of the electronic excitation process because of their electrochemical doping features and anisotropic electronic excitation properties. Depolarised Raman scattering was acquired, and additional electronic excitation was evident in the Raman active mode. Thus, the evaluation of the singularities and electronic states raised the possibility that the excitation due to different energies caused the anisotropic polarised properties.

Next, the strong-coupling regime was achieved using LSP and molecular excitons, which induced the modulation of the electronic state (Chap. 5). The system was controlled by changing the LSP energy against the dye exciton. The resonant condition between LSP and dye exciton energy shows a hybridised state and energy separation. The hybridised state was suggested for use in the new electronic excitation state. A highly polarised SERS signal was acquired in the hybridised system, and Raman intensity was enhanced at the resonant condition between the hybrid states and the excitation. This shows that the molecules in the confined EM field provided the hybridised state. In addition, the energy state of strong coupling was tuned in the electrochemical environment (This chapter). By tuning the electrochemical potential, LSP energy was controlled associated with the free electron density of nanostructure. While features of the environment of the strong-coupling system were maintained, such as nanostructure size and the number of dye layers, the coupling strength was tuned. The developed hybridised state was well controlled, showing that the electronic excitation processes were also controlled.

In conclusion, the present investigations on a newly developed electronic state gathered indispensable information about the photoexcitation process at the interface between metal and adsorbed molecules. The alternation of the excited state is expected to realise a new process of photochemical reaction on the metal surface. This possibility opens the door for us to develop the efficiency, selectivity and sustainability of various photochemical reactions.

References

1. N. Kometani, M. Tsubonishi, T. Fujita, K. Asami, Y. Yonezawa, Langmuir **17**, 578–580 (2001)
2. S. Franzen, J.C.W. Folmer, W.R. Glomm, R. O'Neal, J. Phys. Chem. A **106**, 6533–6540 (2002)
3. G.A. Wurtz, P.R. Evans, W. Hendren, R. Atkinson, W. Dickson, R.J. Pollard, A.V. Zayats, W. Harrison, C. Bower, Nano Lett. **7**, 1297–1303 (2007)
4. N.T. Fofang, T.H. Park, O. Neumann, N.A. Mirin, P. Nordlander, N.J. Halas, Nano Lett. **8**, 3481–3487 (2008)
5. W.H. Ni, Z. Yang, H.J. Chen, L. Li, J.F. Wang, J. Am. Chem. Soc. **130**, 6692 (2008)
6. N. Cade, T. Ritman-Meer, D. Richards, Phys. Rev. B **79**, 241404 (2009)
7. W. Ni, T. Ambjornsson, S.P. Apell, H. Chen, J. Wang, Nano Lett. **10**, 77–84 (2010)
8. Y.-W. Hao, H.-Y. Wang, Y. Jiang, Q.-D. Chen, K. Ueno, W.-Q. Wang, H. Misawa, H.-B. Sun, Angew. Chem. **123**, 7970–7974 (2011)
9. N.T. Fofang, N.K. Grady, Z. Fan, A.O. Govorov, N.J. Halas, Nano Lett. **11**, 1556–1560 (2011)
10. C. Bonnand, J. Bellessa, J. Plenet, Phys. Rev. B **73**, 266808 (2006)
11. S. Balci, C. Kocabas, S. Ates, E. Karademir, O. Salihoglu, A. Aydinli, Phys. Rev. B **86**, 235402 (2012)
12. J. Dintinger, S. Klein, F. Bustos, W. Barnes, T. Ebbesen, Phys. Rev. B **71**, 035424 (2005)
13. P. Vasa, R. Pomraenke, G. Cirmi, E. De Re, W. Wang, S. Schwieger, D. Leipold, E. Runge, G. Cerullo, C. Lienau, ACS Nano **4**, 7559–7565 (2010)
14. A. Salomon, R.J. Gordon, Y. Prior, T. Seideman, M. Sukharev, Phys. Rev. Lett. **109**, 073002 (2012)
15. J. Bellessa, C. Bonnand, J. Plenet, J. Mugnier, Phys. Rev. Lett. **93**, 036404 (2004)
16. T.K. Hakala, J.J. Toppari, A. Kuzyk, M. Pettersson, H. Tikkanen, H. Kunttu, P. Torma, Phys. Rev. Lett. **103**, 053602 (2009)
17. C. Symonds, C. Bonnand, J.C. Plenet, A. Brehier, R. Parashkov, J.S. Lauret, E. Deleporte, J. Bellessa, New J. Phys. **10**, 065017 (2008)
18. Fontcuberta i Morral, A.; Stellacci, F. *Nature Mater.* **2012,** 11, 272–273
19. J.A. Hutchison, A. Liscio, T. Schwartz, A. Canaguier-Durand, C. Genet, V. Palermo, P. Samori, T.W. Ebbesen, Adv. Mater. **25**, 2481–2485 (2013)
20. A. Canaguier-Durand, E. Devaux, J. George, Y. Pang, J.A. Hutchison, T. Schwartz, C. Genet, N. Wilhelms, J.-M. Lehn, T.W. Ebbesen, Angew. Chem. **125**, 10727–10730 (2013)
21. T. Schwartz, J.A. Hutchison, C. Genet, T.W. Ebbesen, Phys. Rev. Lett. **106**, 196405 (2011)
22. A.L. Baudrion, A. Perron, A. Veltri, A. Bouhelier, P.M. Adam, R. Bachelot, Nano Lett. **13**, 282–286 (2013)
23. A. Berrier, R. Cools, C. Arnold, P. Offermans, M. Crego-Calama, S.H. Brongersma, J. Gomez-Rivas, ACS Nano **5**, 6226–6232 (2011)
24. A.C. Templeton, J.J. Pietron, R.W. Murray, P. Mulvaney, J. Phys. Chem. B **104**, 564–570 (1999)
25. P. Mulvaney, Metal nanoparticles: double layers, optical properties, and electrochemistry, in *Nanoscale Materials in Chemistry* (Wiley, 2002), pp. 121–167
26. A.H. Ali, R.J. Luther, C.A. Foss Jr., G.B. Chapman, Nanostruct. Mater. **9**, 559–562 (1997)
27. X. Wu, S.K. Gray, M. Pelton, Opt. Express **18**, 23633–23645 (2010)
28. C.M.S. Yau, S.I. Pascu, S.A. Odom, J.E. Warren, E.J.F. Klotz, M.J. Frampton, C.C. Williams, V. Coropceanu, M.K. Kuimova, D. Phillips, S. Barlow, J.-L. Bredas, S.R. Marder, V. Millar, H.L. Anderson, *Chem. Commun.* 2897–2899 (2008)
29. M. Moskovits, Rev. Mod. Phys. **57**, 783–826 (1985)
30. M. Inoue, K. Ohtaka, J. Phys. Soc. Jpn. **52**, 3853–3864 (1983)
31. B. Pettinger, J. Chem. Phys. **85**, 7442–7451 (1986)
32. P. Johansson, H. Xu, M. Käll, Phys. Rev. B **72**, 035427 (2005)

33. K.-I. Yoshida, T. Itoh, H. Tamaru, V. Biju, M. Ishikawa, Y. Ozaki, Phys. Rev. B **81**, 115406 (2010)
34. C.M. Galloway, P.G. Etchegoin, E.C. Le Ru, Phys. Rev. Lett. 103, 063003 (2009)
35. P.G. Etchegoin, E.C. Le Ru, *Anal. Chem.* 82, 2888–2892 (2010)
36. T. Itoh, M. Iga, H. Tamaru, K.-I. Yoshida, V. Biju, M. Ishikawa, J. Chem. Phys. **136**, 024703 (2012)
37. M. Ringler, A. Schwemer, M. Wunderlich, A. Nichtl, K. Kürzinger, T.A. Klar, J. Feldmann, Phys. Rev. Lett. **100**, 203002 (2008)
38. Y. Chen, K. Munechika, D.S. Ginger, Nano Lett. **7**, 690–696 (2007)
39. B. Takimoto, H. Nabika, K. Murakoshi, J. Phys. Chem. C **113**, 11751–11755 (2009)

Curriculam Vitae

1. Personal Information

Name:	Fumika Nagasawa
Nationality:	Japanese
Gender:	Female
Current affiliation:	Kanae Co., Ltd.

2. Education

Mar. 2009	Bachelor of Science Department of Chemistry, Faculty of Science, Hokkaido University, Sapporo, Japan
Mar. 2011	Master of Science Department of Chemistry, Faculty of Science, Hokkaido University, Sapporo, Japan
Mar. 2014	Doctor of Philosophy in Science Graduate School of Chemical Sciences and Engineering, Hokkaido University, Sapporo, Japan

3. Professional Carrier

Apr. 2011–Mar. 2014	Research Fellow DC1 Japan Society for the Promotion of Science
Apr. 2014–Mar. 2016	Kaneka Co., Ltd., Researcher
Apr. 2016–Present	Kanae Co., Ltd., Researcher

4. Publication

4.1. Papers

1. Hiro Minamimoto, Fumiya Kato, Fumika Nagasawa, Mai, Takase, and Kei Murakoshi, "Electrochemical Control of Strong Coupling States between Localized Surface Plasmons and Molecule Excitons for Raman Enhancement", *Faraday Discussion*, doi:10.1039/C7FD00126F (2017).

F. Nagasawa, *Studies on the Plasmon-Induced Photoexcitation Processes of Molecules on Metal Surfaces*, Springer Theses,
DOI 10.1007/978-4-431-56579-6

2. Yuko Yonezawa, Hiro Minamimoto, Fumika Nagasawa, Mai Takase, Satoshi Yasuda, and Kei Murakoshi, "In-Situ Electrochemical Surface-Enhanced Raman Scattering Observation of Molecules Accelerating the Hydrogen Evolution Reaction", *J. Electroanal. Chem.*, doi:10.1016/J.HELECHEM.2017.04.049 (2017).
3. T. Suzuki, Y. Uchimura, F. Nagasawa, T. Takeda, H. Kawai, R. Katoono, K. Fujiwara, K. Murakoshi, T. Fukushima, A. Nagaki, and J. Yoshida, "Expandability of Ultralong C–C Bonds: Largely Different C1–C2 Bond Lengths Determined by Low-Temperature X-Ray Structural Analyses on Pseudopolymorphs of 1,1-Bis(4-fluorophenyl)-2,2-bis(4-methoxyphenyl)pyracene", *Chem. Lett.*, **43**, 86–88 (2014).
4. F. Nagasawa, M. Takase, and K. Murakoshi, "Raman Enhancement via Polariton States Produced by Strong Coupling between a Localized Surface Plasmon and Dye Excitons at Metal Nanogaps", *J. Phys. Chem. Lett.*, **5**, 14–19 (2014).
5. T. Shoji, M. Shibata, N. Kitamura, F. Nagasawa, M. Takase, K. Murakoshi, A. Nobuhiro, Y. Mizumoto, H. Ishihara, and Y. Tsuboi, "Reversible Photoinduced Formation and Manipulation of a Two-Dimensional Closely Packed Assembly of Polystyrene Nanospheres on a Metallic Nanostructure", *J. Phys. Chem. C*, **117**, 2500–2506 (2013).
6. T. Shoji, J. Saitoh, N. Kitamura, F. Nagasawa, K. Murakoshi, H. Yamauchi, S. Ito, H. Miyasaka, H. Ishihara, and Y. Tsuboi, "Permanent Fixing or Reversible Trapping and Release of DNA Micropatterns on a Gold Nanostructure Using Continuous-Wave or Femtosecond-Pulsed Near-Infrared Laser Light", *J. Am. Chem. Soc.*, **135**, 6643–6648 (2013).
7. T. Konishi, M. Kiguchi, M. Takase, F. Nagasawa, H. Nabika, K. Ikeda, K. Uosaki, K. Ueno, H. Misawa, and K. Murakoshi, "Single Molecule Dynamics at a Mechanically Controllable Break Junction in Solution at Room Temperature", *J. Am. Chem. Soc.*, **135**, 1009–1014 (2013).
8. F. Nagasawa, M. Takase, H. Nabika, and K. Murakoshi, "Polarization Characteristics of Surface-Enhanced Raman Scattering from a Small Number of Molecules at the Gap of a Metal Nano-Dimer", *Chem. Commun.*, **47**, 4514–4516 (2011).
9. F. Nagasawa, M. Takase, H. Nabika, and K. Murakoshi, "Characteristic Surface-Enhanced Raman Scattering from a Small Number of Molecules in an Anisotropic Electromagnetic Field at Metal Nano-Gap", *Trans. Mater. Res. Soc. Jpn.*, **35**, 279–282 (2010).

4.2. Review and Book Chapter

1. F. Nagasawa, M. Takase, H. Nabika, and K. Murakoshi, "Single Molecule Surface-Enhanced Raman Scattering as a Probe for Adsorption Dynamics on Metal Surfaces", in *Vibrational Spectroscopy at Electrified Interfaces*, edited by C. Korzeniewski, B. Braunschweig and A. Wieckowski (John Wiley & Sons, New York, 2013), pp. 220–237.

2. M. Takase, F. Nagasawa, H. Nabika, and K. Murakoshi, "Depolarization of Surface-Enhanced Raman Scattering Photons from a Small Number of Molecules on Metal Surfaces", in *Frontiers of Surface-Enhanced Raman Scattering: Single-Nanoparticles and Single Cells*, edited by Y. Ozaki, K. Kneipp and R. R. Aroca (John Wiley & Sons, New York, 2014), pp. 89–106.
3. F. Nagasawa, "In-situ Observation of Electrified Interfaces using Metal Nanostructure" (in Japanese), *Analyst*, **10**, 663 (2013).
4. H. Nabika, M. Takase, F. Nagasawa, and K. Murakoshi, "Toward Plasmon-Induced Photoexcitation of Molecules", *J. Phys. Chem. Lett.*, **1**, 2470–2487 (2010).

5. Awards

1. Best Poster Award, International Conference on Nanoscopic Colloid and Surface Science, Sep. 2010
2. Lilac Poster Award, Hokkaido Branch, Electrochemical Society of Japan, Jun. 2011
3. Best Poster Prize, International Society of Electrochemistry, Aug. 2012
4. ICAVS-7 Poster Award, 7th International Conference on Advanced Vibrational Spectroscopy, Aug. 2013
5. Otsuka Award for Outstanding Thesis, Hokkaido University, Mar. 2014.